Acclaim for Henry Petroski's

The Book on the Bookshelf

"The charm of this book lies in the way it helps us take a fresh look at an old, long-familiar object, seeing it, as its various inventors and innovators did, for the first time." —*The Christian Science Monitor*

"[Petroski's] latest book will surely find its way onto the bookshelves of his many fans. . . . [It is] generously peppered with excellent, indispensable illustrations." —*The Miami Herald*

"Petroski has triumphantly revealed the surprises lurking beneath the seemingly self-evident. . . . With his characteristic ability to balance practicality with genial connoisseurship, [he] is a well-equipped guide to the subject." —*Newsday*

"An informative and delightful book." —*The Raleigh News & Observer*

"A tightly focused overview of how, over the past two millennia, scrolls and bound volumes have been housed, arranged and cared for. . . . Fascinating." —*The Washington Post Book World*

"Rich and rewarding." —*The Boston Book Review*

HENRY PETROSKI

The Book on the Bookshelf

Henry Petroski is the Aleksander S. Vesić Professor of Civil Engineering and Professor of History at Duke University. He is the author of eight previous books.

ALSO BY HENRY PETROSKI

Remaking the World

Invention by Design

Engineers of Dreams

Design Paradigms

The Evolution of Useful Things

The Pencil

Beyond Engineering

To Engineer Is Human

The Book on the Bookshelf

by

HENRY PETROSKI

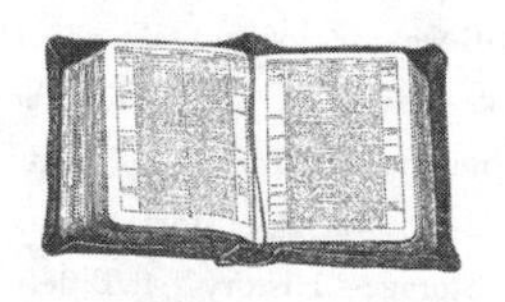

VINTAGE BOOKS
A DIVISION OF RANDOM HOUSE, INC.
NEW YORK

FIRST VINTAGE BOOKS EDITION, SEPTEMBER 2000

 Published in the United States by Vintage Books, a division of Random House, Inc., New York, and simultaneously in Canada by Random House of Canada Limited, Toronto. Originally published in hardcover in the United States by Alfred A. Knopf, a division of Random House, Inc., New York, in 1999.

The Library of Congress has cataloged the Knopf edition as follows:
Petroski, Henry.
The book on the bookshelf / by Henry Petroski. —1st ed.
p. cm.
Includes bibliographical references and index.
ISBN 0-375-40649-2 (alk. paper)
1. Shelving for books—History. 2. Shelving for books—Europe—History. 3. Bookbinding—History. 4. Bookbinding—Europe—History. 5. Books—Storage—History. 6. Books—Europe—Storage—History. I. Title.
Z685.P48 1999
022'.4'09—dc21 99-14336
CIP

Vintage ISBN: 0-375-70639-9

Author photograph © Catherine Petroski
Book design by Robert C. Olsson

www.vintagebooks.com

Printed in the United States of America
10 9 8 7 6 5 4 3 2 1

To Karen and Jason,
whose bookshelves are full

Contents

Preface

One evening, while reading in my study, I looked up from my book and saw my bookshelves in a new and different light. Instead of being just places on which to store books, the shelves themselves intrigued me as artifacts in their own right, and I wondered how they came to be as they are. Question led to question, and I began to look for answers in—where else?—books. Books led me to libraries, where I naturally encountered more bookshelves. I have found that, as simple as the bookshelf might appear to be as an object of construction and utility, the story of its development, which is intertwined with that of the book itself, is curious, mysterious, and fascinating.

The books that have helped me understand and tell the story of the bookshelf are acknowledged in this one's bibliography. Libraries, librarians, and library staff members who helped me assemble the bibliography should be acknowledged here. As I have found time after time, the libraries of Duke University are an indispensable resource for me, not only for their own wonderful collections but also for the access that they provide through interlibrary loan to the rest of the library world. I am grateful for the continuing carrel assignment I have enjoyed in Perkins, Duke's main library, and for the helpfulness of all its staff. My main point of contact with the Duke libraries is the Aleksandar S. Vesić Engineering Library, and its staff, Dianne Himler, Tara Bowens, and librarian Linda Martinez. Their patience with what must seem my endless and capricious requests has my heartfelt appreciation. I am also grateful to Eric Smith, a tireless reference librarian in Perkins, and to the university librarian, David Ferriero, who provided me with letters of introduction to British libraries and who referred me to Janet Chase of the Library of Congress. It was she who not only arranged for me a tour of that library's historic stacks, which are closed even to most library staff, but also thoughtfully arranged for a parking space right in front of the Jefferson Building. Joseph Puccio, public service officer at the Library of Congress, provided a most thorough and informative tour of the stacks that revolutionized library storage.

I am also grateful to the many other librarians and library staff for the time and freedom they gave me to explore their bookstacks and the books in their collections. Among the visits most helpful to my arcane purposes were those to Yale University's Beinecke Rare Book & Manuscript Library and Sterling Memorial Library; the University of Iowa's rare-book library and its distinguished collection in the history of hydraulics; and the Smithsonian Institution's Dibner Library of the History of Science and Technology, where Leslie Overstreet, reference librarian in the Smithsonian Institution Libraries, was especially helpful in showing me around the Dibner collection. I am also grateful to Alison Sproston, sub-librarian in Trinity College, Cambridge University, for guiding my visit to the Wren Library; to Richard Luckett and Aude Fitzsimons, librarian and assistant librarian in Magdalene College, Cambridge, for enabling me to see the Pepys Library on short notice; and to Dan Lewis and Alan Jutzi of The Huntington for their guided tour of that library's rare book stacks and vault.

Ashbel Green, my editor, Asya Muchnick, his assistant, Melvin Rosenthal, the production editor, Robert Olsson, the designer, and everyone else who participated in the production of this book from my manuscript, have once again demonstrated the excellence of Alfred A. Knopf, for which I am enormously grateful.

My children, Karen and Stephen, no longer living at home, have participated less directly in this book than they have in some previous ones, but the questions I expected they might ask and the observations I thought they might make have informed this book no less than my others. My wife, Catherine, has once again provided a reality check on my ideas and a fair first reading of the manuscript.

H.P.
Durham, North Carolina,
and Arrowsic, Maine

The BOOK on the BOOKSHELF

CHAPTER ONE

Books on Bookshelves

My reading chair faces my bookshelves, and I see them every time I look up from the page. When I say that I see them, I speak metaphorically, of course, for how often do we really *see* what we look at day in and day out? In the case of my bookshelves, in fact, I tend to see the books and not the shelves. If I think consciously about it and refocus my eyes—the way I must do when viewing optical illusions, to see the stairs go up instead of down or the cube recede in perspective to the right rather than to the left—I can see the shelves, but usually only their edges and maybe the bottoms of the upper shelves, and seldom the shelves complete and the shelves alone. Even when the bookshelves are bare, I tend to see not the shelves themselves but the absence of books, for the shelves are defined by their purpose.

If the truth be told, neither do I see the books without the shelves. The bottoms of the books rest squarely on the shelves, and the rows of books are aligned against gravity. The tops of these same books present a ragged line, of course, but even this is defined by the shelf on which they rest, and is emphasized by the straight edge of the shelf above. Books and bookshelves are a technological system, each component of which influences how we view the other. Since we interact with books and bookshelves, we too become part of the system. This alters our view of it and its components and influences our very interaction with it. Such is the nature of technology and its artifacts.

An attempt to gain perspective on the bookshelf is not a simple matter. The bookshelves in my study go from floor to ceiling and nearly cover one of its walls, but because my study is not grand, I cannot easily distance myself from the wall of shelves. Even when I first moved into this study, when it and the bookshelves were bare, I could not stand back far enough to view the shelves entire. No matter where I stand before this wall of shelves, I see the bottoms of some and the tops of others, the left side of some of the vertical supports and the right side of others. I never see all of a single shelf at a single time. I can, of course, take it for granted that all the shelves are identical and so infer that when I see the bottom of one shelf I see the bottom of all shelves, but there is something not wholly satisfying about such philosophizing, common as it is.

While reading in my chair late one evening I perceived, for whatever reason, the bookshelf beneath a row of books in a new light. I saw it as a piece of infrastructure, taken for granted if not neglected, like a bridge beneath a line of cars, and I wanted to know more about the nature and origins of this ubiquitous thing. But where to begin? Was it meaningful to ask why the bookshelf is horizontal and why books are placed vertically upon it? Or are these facts so obvious as to need no explanation? Going further, was there anything to be gained by asking why we shelve our books with their spines facing outward, or is this simply the only logical way to shelve them? Don't books go on bookshelves, as nuts go on bolts, only one way?

As it turns out, the story of the bookshelf is rooted in the story of the book, and vice versa. It may be strictly true that books can exist without bookshelves, and we can imagine the Library of Congress or even the local public library with books contained in boxes, stacked on the floor, or stored in piles like firewood or coal. The bookshelf, however, can hardly be imagined without the existence of books. That is not to say that without books we would not have shelves, but they would certainly not be *book*shelves. The bookshelf, like the book, has become an integral part of civilization as we know it, its presence in a home practically defining what it means to be civilized, educated, and refined. Indeed, the presence of bookshelves greatly influences our behavior.

Authors often have their pictures taken in front of bookshelves, but why? Certainly they have not written all the books before which they stand. Perhaps they want to show us how many books they have read in order to write theirs, and that we will not have to read if we delve into their comprehensive essay or historical novel, with its extensive notes or wide-ranging bibliography, explicit or implicit.

Since the book on which their photo appears is seldom, if ever, on the bookshelf behind them, perhaps these authors are sending the subliminal message that we should go to the bookstore and buy their book to complete the shelf.

But can a bookshelf ever be complete? There are well over fifty thousand books published every year in America alone. Can anyone read that many books even in a lifetime? The math is not hard to do. If we read roughly a book a day, we can read about one thousand books every three years. Assuming that we start when we are four years old and live to the ripe old age of ninety-four, we could then read about thirty thousand books in a lifetime. What would it take to shelve that many books? Assuming each book requires on average an inch of shelf length, we would need about 2,500 feet of shelving. It would take a house with six or seven large rooms fitted with bookshelves on every wall to hold that many books, which would make it not a home but a bookstore—or a small town's public library.

But if we walked into such a house, would we see books or bookshelves? Which in fact do we see when we walk into a library? In virtually all cases, the books are the focus of our attention. Like the steps beneath a group of people being photographed, the shelves go largely unseen; they are there but not there. They are the infrastructure. Yet the bookshelf is also conspicuous in its absence. When we enter a living room without books or bookshelves, we wonder if the people in the house do nothing but watch television.

Ironically, the bookshelf is a familiar prop on television, being frequently part of the deep background for interviews on shows ranging from *Today* to *Nightline.* Congressmen and senators hold news conferences carried on C-SPAN before a bookcase that is no wider than the camera's frame. (Are the books in it real?) Whenever Newt Gingrich wore his bookshelf necktie on the set, he was faced and backed by books. Lawyers and professors are often interviewed in front of shelves of books, the producer apparently wanting to associate the experts with the authority of the library behind them.

A propped-up prop, the bookshelf plays a supporting role to the book. It is not only the backdrop but the stage itself upon which books line up for applause. And yet, important as its role has been in the history of civilization, the bookshelf seldom even gets mentioned in the program; it is treated as a supernumerary, taken for granted, and ignored. There is much anecdotal evidence that this is the case.

During a cocktail party, the wife of a colleague retreated to my study to nurse her newborn baby. When she emerged some time later with the sleeping child, she said she hoped I didn't mind that she had browsed through my bookshelves, but she was interested to find several books that she had fond memories of reading. It is, of course, unremarkable that she had nary a word to say about the shelves on which the books reside. But when another guest visited my study on a different occasion, his focus on the books to the exclusion of the shelves was notable.

On a pleasant spring afternoon, this guest browsed in my study while I looked for something to give him to read on the plane. His browsing soon became perusing, and he looked at the books on my bookshelves with an intensity of purpose that was not unfamiliar. It is a spectator sport to look at someone else's books, if not an act of voyeurism or armchair psychology. My guest seemed hardly to overlook a title on the shelves, and he remarked to me that he always found it interesting to see what people owned and read. Naturally, he would be interested: my guest was a psychologist, a specialist in cognitive science who has served as a consultant on computer-user interfaces, and who now advised a major office-equipment manufacturer about what products to develop and what features to design into them. He had written with insight on the design of everyday things, paying special attention to their use. As a reader of his books, I did not think he would miss a thing, no matter what he was looking at.

Earlier in the day I had shown him around town. We had stopped at a new public policy studies building which has been lauded for the attention its architect paid to how the structure and its space would be used. As soon as we entered, we could see that this was not a conventional building. Numerous offices and conference rooms open onto balcony-like hallways that overlook two sides of a common space which is only loosely bordered on its other two sides by tiers of open lounge areas, which also overlook and help define the atrium. It does not appear that a person can move from one part of the building to another without walking along a hall or stairway that is open to the common space, where chance encounters among those in the building are likely to be frequent, as no doubt planned. The arrangement reminded me of the National Humanities Center, where people enter and leave through the commons, which also serves as the dining area where visiting scholars writing books on everything from pencils to phenomenology gather to converse. My guest was immediately struck by the thoughtful design of the new building, noting subtle details that go unnoticed by most of us, like lighting fixtures over bulletin boards and the hardware on the doors, which he has written about with special insight and passion. Having begun to think about this book, I was hoping to see the bookshelves in the building's offices, but no offices were open on the Saturday afternoon that we visited.

Back in my study, we talked not about things or even about books as objects but about the ideas they contain and how different categories of books were grouped on my bookshelves. My guest found and commented on some familiar titles that he no doubt expected, like Tracy Kidder's *The Soul of a New Machine* and many books about bridges, but he expressed surprise at finding some others there. I explained that many of those having to do with the design of computer software had been sent or given to me by readers of my own books on the design of bridges and other useful things. Since I have argued that design is design, regardless of the object being designed, the collection of books represents to me a unity of theme, if not a downright obsession with a few ideas, but I admitted difficulty in deciding exactly where to shelve a given book that touches on more than one aspect of the theme. My guest surely formed an opinion about what I read and how I work in my study, but the conversation turned to computers and the features I should look for in a laptop, since I told him I was in the market for one.

If my guest formed an opinion about me through the books on my bookshelves, this confirmed one of my current hypotheses: that for all the attention even the most observant of us pays to useful things, we all but ignore the infrastructure upon which they rest. My guest made no comment about the bookshelves, even after I tried to steer the conversation in that direction. Even the fact that the topmost of the floor-to-ceiling shelves were out of his reach did not elicit a comment from him, this critic of everything from the design of telephone systems to the location of electric-light switches. "The dust and silence of the upper shelf," about which Lord Macaulay wrote, remained undisturbed in our conversation as well. Once in place and with books upon it, the bookshelf has no moving parts and no obvious function except to stay where it is and support a line of books. It is like a common bridge on a small country road, there but not there to all who use it every day. Yet let the bridge be washed out in a flood, and suddenly it becomes the most important topic of discussion in the county. So it is with technology generally: it is most present in its absence.

When I began working on this book, I saw bookshelves where I once saw books, but not everyone shared my perspective. At a dinner party one evening, in the home of a historian who built his own wall of bookcases sized just right to hold the many paperbacks that historians are wont to have, I commented on the bookshelves I had all but ignored on previous visits. The conversation eventually turned from the pride of workmanship in making such things to the more general topic of books and their arrangement on the shelves. Since I was full of thoughts of how books were shelved in medieval times and the evolution of the bookshelf as we know it, I tried to steer the after-dinner talk back to bookshelves. I was interested to learn that their origins are not widely known even among historians, especially those whose period is not the Middle Ages. Talking to a retired English professor some months later, I found again that the physical nature of medieval books and the fact that they were chained to the bookshelf is not common knowledge among students of books whose specialties lie in later centuries.

It was not only from scholars but also from librarians that I discovered that neither the history of books and their care nor the design and development of the furniture in which they are stored and displayed are widely known. The title alone of an early book that I consulted, *The Chained Library,* by Burnett Hillman Streeter, aroused curiosity among librarians and library clerks alike when I requested it. The book, published in 1931, seems to have been regularly if infrequently checked out for the first decade it was in the library, but the last due date stamped on the slip in the back of the book is OCT 28 '41. Judging from the signatures on the charge card, still in its pocket on the inside back cover, the book may have been read by no more than ten people at one of the outstanding research libraries in the country. There is no record in the book of its having been checked out for the next decade, at least. I cannot know what happened beyond that, for the library's circulation procedures changed in the early 1950s. From that time on, the date-due stamp and charge card were abandoned in the back of the book, artifacts of an earlier time when anyone whose signature was on the card could be expected to be known by the librarian. In any case, I found out that the subject of *The Chained*

Library, like the older procedures for keeping circulation records, was generally not known to younger librarians. They did not share my interest in the history of libraries, or at least not that of their furniture and ways.

After I had read *The Chained Library,* and before it John Willis Clark's *The Care of Books,* the seminal work on the subject, I found myself visiting Yale University and the Beinecke, one of the premier rare-book libraries in the world. I was graciously shown around by a very enthusiastic and helpful guide, but when I asked if the library held any books that still retained traces of the hardware by which they were once chained, he could not answer. However, with the help of the computerized catalog, a library clerk was able to search for the word "chain" among all the descriptions of the library's holdings. There were plenty of items relating to the chain stitching used in old bindings, and there were also a few books in the collection that still bore the marks of where a chain had once been attached to the leather-covered and boss-studded wooden board in which the book was bound. According to the computer, there was also at least one book that has part of a real chain still attached, and I asked to see it. The book is kept in a custom-made box that holds the several heavy black chain links in a compartment separate from the one that holds the book, thus protecting the leather from abrasion by the wrought iron. The artifact seemed to be as curious to those behind the service counter as it was to me, further confirming my growing confidence that the story of the chained book, which is central to the story of the bookshelf, was one that had to be told, not only for its inherent interest but also for its relevance as a case study—a bookcase study—in the evolution of an artifact, a tool for explaining how technology is embedded in and shapes our culture.

It is understandable that most of us think more about books than about bookshelves, but there have been those who have given the infrastructure its due. Henry Cuyler Bunner, longtime editor of the humor magazine *Puck,* went so far as to write:

> *I have a bookcase, which is what*
> *Many much better men have not.*
> *There are no books inside, for books,*
> *I am afraid might spoil its looks.*

Though books may indeed spoil some bookcases, sometimes it is the bookcases that are rough on the books and almost discourage their use. When I moved into my present office at Duke, it was already equipped with bookcases that were handsome and adjustable in shelf height. Because the bookcases are made of heavy particle board beneath their walnut veneer and because the shelves are deep but not very long, they are sturdy and show no visible sag under the heaviest load of books. The bookcases are, however, not very tall, so I adjusted the shelf supports to accommodate the maximum number of shelves at the necessary heights to hold the variety of book sizes that I wanted to have in the office. As a result, I have my books grouped largely by height, and there is little room left above the books on any of the shelves. Indeed, on some tightly packed shelves there is hardly room to get a firm grip on a book to remove it from the shelf. According to one guide on how to care for books, the test of whether they are packed too tightly on the shelves is this: "Can you grasp the sides of a book with your index and middle fingers and your thumb, and gently remove it from the shelf without dislodging the books on either side?" In my case, I cannot, and I must follow the good advice in *Martha Stewart Living:* "To retrieve a book, push in the books at either side and pull gently."

A common procedure, when there is room of course, is to place one's finger on the top of a book and pull gently against the headband to rotate the book in its place until its top corner projects out enough from the other books on the shelf for it to be grasped and removed. *Martha Stewart Living* does not approve: "Never hook your finger over the top of the spine." The problem with doing so is that, when the books are too tightly squeezed together on the shelf, it can lead to broken fingernails or, perhaps worse, to torn book bindings. As a nineteenth-century "handy-book" warned, "Never pull a book from the shelf by the head-band; do not toast them over the fire, or sit on them, for 'Books are kind friends, we benefit by their advice, and they reveal no confidences.' "

Books and bookshelves were viewed more mechanically by the inventor Charles Coley, of Culver City, California, who looked into the matter of taking books off the shelf and found "no prior art which attempts to solve the problem." In 1977 he received a patent on a "book ejection apparatus," which consists of a springboard-like device that fits against the entire back wall of the bookcase and "utilizes an action-reaction" principle to eject a book from its position. To remove a book, one counterintuitively pushes the book against the bookshelf's back wall. This compresses the springs behind the board, and the spring force pushes the desired volume out from between its neighbors. (The device is similar to those hidden cabinet-door latches whereby one pushes in on the door to get it to open out.) Like many inventions, Coley's might take some getting used to in order to operate properly—which it might not ever do anyway if the books are too tightly packed into the bookcase.

Putting a book back on the shelf in such circumstances can be as difficult as putting a sardine back in a can. A bookshelf appears to abhor a vacuum, and so the void that is created when one book is removed is seldom adequate to receive the book again. Like a used air mattress or roadmap, which can never seem to be folded back into the shape in which it came, the book opened seems to have a new dimension when reclosed. Where it once fit it no longer does, and it has to be used as a wedge to pry apart its formerly tolerant neighbors in order to get a foothold on the shelf. Invariably, the book I push back into its place scrapes along its neighbors and pushes them back a little. Where there is ample room above them, the disturbed books can be realigned with a little effort. However, in my office, where I cannot easily reach in to pull the books back out and align their spines, I find myself pushing the whole shelf back a bit to realign them. I cannot just push the books all the way to the rear of the shelf, of course, because they do not all have the same width, and so the shelf of them would present a rather ragged appearance. In time, however, so many of the books end up pushed all the way back that I have to take a whole section of them out and reposition them near the front edge of the shelf.

The arrangement of my books deep in the bookcase had not bothered me, because I generally preferred to keep them a couple of inches back from the front edge of the shelf anyway. When or why I began to do this, I cannot say, but I do not recall, before experimenting while writing this book, that I ever pulled my books all the way to the front of the shelf, unless the deepest book in the line was as wide as the shelf was deep, thus necessitating all the other books to be pulled forward if there were to be any alignment at all. Having a few inches of shelf in front of the books seemed natural and desirable to me, to balance the few inches of shelf in the back. The books were thus centered from front to back, putting nearly equal loads on the shelf supports, which seemed structurally neat and appropriate. Unlike in an institutional library, where the shelves are often facing narrow aisles and where at the ends of the shelf books pushed back from the edge might not easily be seen as one walks down the aisle, in my home and office the bookshelves are against the wall, with more than an aisle's width to stand away and view them head on. I find books pulled forward to the edge of a shelf make the bookcase look top-heavy and a bit tight for them, like a suit one has outgrown. Books at the edge also make my bookcases look two-dimensional, without depth, somewhat like wallpaper. Where bookshelf headroom exists, there is some depth provided by the ragged space above the books, of course, but having shadows before the books in addition gives them a more cradled look.

The recessed alignment of books on my bookshelves also gives me a narrow shelf space before the books where I can keep mementos like pencils and letter openers. It all seemed sensible to me until one day a writer visiting my office expressed surprise at my arrangement and remarked that he always pulled his books forward to the edge of his bookshelves and thought that was the proper way to display them. I did not have a definitive answer for him at the time, and I still do not, but I have since learned that the literary critic Alfred Kazin kept his books well back from the edge of the shelf, giving him room to display photos of his grandchildren and to lay down books being read. Like so many questions of design and the human adaptation to and interface with technology, there are arguments that can be presented in support of either option. I was, however, pleased to have the visitor question my book arrangement, for it assured me that I was not alone in thinking about bookcases and how to use them. But how do we proceed in thinking about such things?

A book on the bookshelf is something to be taken down and read; the bookshelf under the book is something to be installed and forgotten. The one object is in service to the other, superior to the other—or such is the conventional wisdom—and the inferior object is something we seldom think about or have reason to. Yet all people and things, common or exalted, have their stories to tell and be told, and these seem more often than not to be gripping tales, full of surprising twists and turns, and always instructive.

What can be more self-evident in form and purpose than the common bookshelf? The idea of a wooden board used for holding books would seem to be at least as old as books themselves, and the fact that the shelf should be flat and horizontal would seem to be dictated by common sense and gravity. Furthermore, that the books on a bookshelf should be vertical—their spines proudly held as straight as a line of cadets—would seem to be another given of the library, whether large or small. It is disconcerting, therefore, to look at Renaissance depictions of scholars in their otherwise neat studies, with many of their books everywhere but on a bookshelf, and those that are shelved in every orientation but straight up and spine outward. Is the vertical book on a horizontal shelf not a law of nature? If not, why not, and how and when did the way we now shelve books come to be an almost universal practice?

The story of the bookshelf cannot be told without telling the story of the book, and how it evolved from scroll to codex to printed volume. These are not arcane subjects that have little relevance for the new millennium; they are among the basic data of civilization that provide a means to a better understanding of the evolving technology of today and to extrapolating it into the future, which will be more like the present and the past than we are usually led to believe.

Looking afresh at and thinking anew about the bookshelf, like looking or thinking from scratch about anything found or made, brings its own rewards, not the least of which are fresh new ways of encountering and experiencing the world. Given that the bookshelf and the books it holds are so dependent upon one another, focusing attention on the normally neglected bookshelf leads us to view the book from a different perspective—from the bottom up, so to speak. When we do look anew at something so familiar as the book, we see a different object, one that has qualities that make it at the same time uniquely distinct from everything else in the world and yet much like so many other things of our experience.

Like wrestlers in a ring, two books on an otherwise empty shelf lean against each other in an uneasy stance. Three books are like one basketball player caught between two opponents in a double-team. More books are like a pack of schoolboys playing Johnny-Ride-a-Pony against the school yard wall. Mostly, however, a shelf-not-full of books is a train full of commuters caught frozen in time leaning against one another in a tenuous balance between gravity and acceleration.

The book on the bookshelf is a curious thing. It cannot easily stand by itself unless it has sufficient girth. The unsupported slim book all too often falls flat on its face or its back with a slap, like the weakling at the beach, embarrassed by its own slightness; the uncontained fat book swells, perhaps with too much pride of purpose or cellulose of notes, as its weighty pages disfigure its spine and spread its covers like the sagging mass of a sumo wrestler squatting before an opponent, daring to be pushed over.

Anne Fadiman, whose *Ex Libris* is a delightful book of essays about books, has written of once losing a twenty-nine-page volume "too slender to bear a title on its vermilion spine" which had become "squashed between two obese shelf-neighbors, much as a flimsy blouse on a wire hanger can disappear for months in an overstuffed closet." In another essay, she revealed her preference for the bookshelf over the closet: "My brother and I were able to fantasize far more extravagantly about our parents' tastes and desires, their aspirations and their vices, by scanning their bookcases than by snooping in their closets. Their selves were on their shelves."

Books spend a lot of time on bookshelves, hanging around near the curb, as it were, waiting for someone to come along with an idea for something to do. Books are the wallflowers at the dance, standing up but leaning on one another and depending upon one another for their collective status. Books are the Martys of Saturday nights, ending up in the same place at the same time week after week. Books in dust jackets are the queue at the bus stop, the line of commuters with their faces hidden in their newspapers. Books are the thugs in the lineup, all fitting a profile but with only one of them expecting to be picked out. Books are the object of searches.

Some books are single-family homes of essays on a theme; some are apartment buildings of anthologies. Books on a shelf are the row houses of Baltimore, the attached homes of Philadelphia, the townhouses of Chicago, the brownstones of New York, all fronted by narrow sidewalks and small backyards that no one but the owner ever sees. The contiguous stepped rooftops present a skyline of sorts, a graph of lives and loves, a cityscape. As in cities every day, passersby walk along the sidewalk on their daily rounds, barely looking at the individual buildings or the people in them. A row of books may hardly be noticed unless we are looking for a title, a call number, a specific address.

Not all books are just one among many, part of the crowd; best-sellers are stars. But no matter how many celebrities or personalities it holds, no matter how many paparazzi trample over it, the bookshelf is a doormat. Bookshelves are the library's infrastructure, the bridges between A and Z on the information back roads, the old county and state routes that the new interstate highway parallels, the trailblazers for the information superhighway. Bookcases are the basic furniture of studies, bookstores, and libraries. The bookshelf is the floor on which books stand; it is the bed on which they sleep until a prince of a reader wakes them up or a talent scout promises to make them a star. Books open up their hearts, but bookshelves simply pine.

What bookshelves wait for, of course, is books. Seldom is a whole shelf of books placed on a bookshelf at once, unless perhaps the person moving the books is one of those jugglers who can catch cigar boxes in midair between other cigar boxes and suspend them—and the audience—with sleight of hand. It could be done with books, but a whole shelf-full? Usually we move a few books at a time or add a new book or two after a birthday party or a visit to the bookstore. The bookshelf is not always full. This might be bliss to librarians, but it is a heart-wrenching void to collectors, for whom the shelf is something to be neither seen nor heard.

The bookcase without a full complement of books is like a daydreaming student's notebook, its lines half filled with substance and half with space. The half-filled bookshelf is also half empty, of course, with books leaning left and right to form M's, N's, V's, and W's to fill the voids between clusters of vertical and not-so-vertical I's.

Though bookshelves are ever ready to lend support from underneath, they cannot always give lateral support to the tottering book. Bookends, those curious constructions that are supposed to hold books back as a dam does water, may or may not support either the slender or the squat. As dams sometimes do, bookends slip and tip over, opening up cracks in the once-tight facade of book backs and allowing blocks of books to topple over in an unsightly pile. It is a manifestation of the eternal video-game conflict between up-and-down and side-to-side motion, the tension between the obelisk and the sledge, each subject to gravity in its own characteristic way. Indeed, gravity, the force that makes bookends work, is the very definition of verticality. Yet it is the equally definitive horizontal force, caused by the bearing down of the bookend's weight, that creates the force which resists sliding.

Contrary to the conventional wisdom, the simplest machine is not the wedge but the block. A Victorian guide to furnishing the home library advised that the "best block" to cause books "to stand upright is made by sawing diagonally in half a cube of wood six inches every way." When called upon, bookends—many of which are really nothing but sculpted blocks—develop a horizontal push to shore up books that want to fall. Friction is the secret, of course, but as with all machines, there is a limit to how hard a bookend can push, because there is a limit to how much friction can be developed between bookend and bookshelf. The heavier and taller the bookend, the better, and the more roughness between the bookend and the shelf, the better. Beyond that, there is little that can be done to a bookend to assist it in its function on the bookshelf.

Some bookends have a thin metal base that fits under the first few books in a line, and the weight of the books then provides the pressure to be converted into friction between bookend and bookshelf. Some bookends of this kind are cleverly and simply made out of a single flat sheet of steel that has been punched out and bent to shape. Such bookends have been very common since they were patented in the 1870s, but they are seldom appropriate for the home library, and they tend not to have the stiffness to withstand the push of heavier books and remain upright. The principle is used with more grace when a handsome piece of wood is employed for the vertical part, and a sturdy piece of metal is securely fastened to its bottom. My wife and I once found some bookends of this kind in a craft store in Indiana. The handsome walnut ends are inlaid with a subtle line of small ceramic tiles, and the base is a piece of heavy plated metal with a thin pad of foam rubber glued to its bottom to increase the friction force that can be developed between it and the shelf. These bookends work very well, remaining vertical and thus keeping the books they contain bolt upright. Unfortunately, nothing is perfect, and the extra thick base that gives the bookend its stiffness raises the last few books in the line a noticeable distance of about ⅛ inch above the shelf. This creates a distracting gap beneath these books, and since they seldom have just the right dimensions so that the book at the edge of the base fits wholly on it, that book straddles the step from the bookend to the bookshelf, and its spine takes on a noticeable skew, because one of its covers is higher than the other. (Bookends that best fit the bookshelf and the books have been made by hollowing out unwanted volumes and filling them with something heavy, but this is anathema to many book lovers. Other bookends are made of hardwood or stone carved to resemble a set of book spines, and these are often least obtrusive.)

Among my most substantial bookends is one that is a 2½-inch section from an actual steel railroad track, ironically the epitome of endlessness. It is by far the weightiest of my bookends, and I have glued a piece of felt to its bottom to keep it from scratching the shelf. This bookend is not pushed aside by even the heaviest line of books. It is, however, sometimes tipped over by taller books, because it is the nature of the rail shape to be top-heavy. I have yet to find the perfect bookend, and I do not expect ever to do so. For every advantage there will be, if not an equal, at least an opposite disadvantage. Such is the nature of made things, and trying to maximize the advantages while minimizing the disadvantages is the object of engineering and of all design.

A board on brackets attached to a wall is certainly frequently used as a bookshelf. Indeed, it is the home improvement store's basic off-the-shelf bookshelf. A set of such shelves generally has no enclosed ends, and so requires bookends in one form or another. Sometimes the brackets for the shelf above can serve as abutments of a kind for the row of books, with a book of just the right height being used to maximum effect. Alternately, books themselves can be employed as gravity bookends, either a particularly wide book that holds back a line of carefully arranged volumes that do not push sideways too much, or a group of books stacked horizontally and providing the weight that the mute machine converts to friction when called upon to do so. As we all know, however, once a long row of books begins to lean over, there is hardly a bookend in the world that can produce enough friction to hold back the torrential surge of bound but unleashed volumes that comes down the shelf.

Bookshelves attached not to a wall by brackets but to the sides of a piece of furniture known as a bookcase may or may not require the assistance of bookends to keep books vertical. When there are enough books to fill the length of a bookcase shelf, bookends become unnecessary, because the ends of the shelves themselves are, literally, bookends, and each book serves as a bookend of sorts for the books it touches, with history shoring up history and novel kissing novel. The

Late-Victorian bookshops sold these bookshelves, made of lightweight wooden boards and wrought-iron rods, designed to be hung on a wall.

bookshelf in a bookcase is thus more than a horizontal board; it is a board with ends. And because they do not rely solely upon the force of friction, those ends can produce much more of a sideways push than bookends. Indeed, as long as the shelves themselves are strong enough to bear up under the weight of books, the inability of a bookcase to contain all the books that can be stuffed into it is a virtually unheard-of event.

If nonslip conditions are welcome under bookends, they are not welcome under books. In my study, my wooden bookcase is painted with cream-colored semigloss paint that is streaked with the colors of the bindings of the previous owner's books, which appear to have been bound predominantly in red and blue. I expect that he, anxious to get his collection out of boxes or off the floor and onto the newly installed and painted shelves, which I suspect that he built and painted himself, did not wait for them to dry thoroughly. The sticky surface provided plenty of adherence between book and shelf to pull or shear a bit of the pigment off of the bindings.

After a friend of mine had arranged his books on his newly varnished shelves, he noticed that some were more difficult to slide off the shelf than others. Among the most troublesome volumes were a few heavy engineering texts, which he reasoned were being hampered from sliding out easily by the friction between the shelf and the bottom of the book. He determined to reduce the friction force by waxing the shelves, as if they were skis, to a high polish, which achieved the desired effect.

The problem of friction between book and shelf was solved in a different way by a professional bookcase designer who used automotive paint on the shelves because it "has enormous impact resistance and allows the books to slide in and out easily." To some book designers, the physical characteristics of the book are more important than its ease of use, and in 1853 the inventor Charles Goodyear had a book printed on rubber sheets and bound in rubber; this volume must have gripped, as a tire does the road, any shelf it rested upon and any books it was shelved beside.

What constitutes a book or a bookshelf depends, like so many things, on definition, and that definition can change with time. Perhaps there is a bibliological analog to the biological one that ontogeny recapitulates phylogeny, or at least a close enough one to warrant the use of this familiar euphonious phrase. Sometimes, especially when we are young, we make our own bookshelves, and sometimes they deviate from the strict horizontal and vertical, but that is not by design. As children we might have devised makeshift bookcases, perhaps by turning an orange crate or other wooden box on its side, and perhaps piling another box on top of it. The slender volumes of children's literature, with footprints smaller than their readers', are notorious for not standing upright by themselves, however, and so children shelve them every which way. But just putting books on a horizontal surface does not necessarily make it a bookshelf. Books on a desk—even books arranged on it ever so neatly straight up between handsome bookends—do not mean a desk is a bookshelf. And books on a windowsill are, well, books on a windowsill.

Ultimately, however, it must be the books it holds that make a board a bookshelf or a box a bookcase. But before any books are added, boards and boxes are merely boards and boxes. As we grow older, our taste in bookshelf-making evolves; many a student has passed through the bricks-and-boards stage. Such shelves have the advantage of being highly transportable as we move from apartment to apartment. In time, however, most of us long for the real thing—bookshelves that were meant to be bookshelves. And, as we advance in our jobs and in our prosperity, we want for our homes the ultimate bookshelves, the built-ins, preferably in an honest-to-goodness study or, better yet, in a room we can call our books' own—a library.

This bookcase, homemade from a wooden box fitted with a shelf and attached bookends with cut-out handles, could be transported with most of the books left shelved.

According to a biography of Edward L. Bernays, the public-relations genius who promoted everything from Dixie cups to Mack trucks and has been called the father of spin, built-in bookshelves began to be popular with architects, contractors, and home decorators in the 1930s, after Bernays was commissioned by publishers who wanted to sell more books. In one version of the story, he got "respected public figures to endorse the importance of books to civilization" and then persuaded those responsible for how civilized homes were furnished to install shelves in them. The homeowner was then expected to buy books to fill up the empty shelves, because Bernays subscribed to the dictum, which he apparently carved out of whole board, that "where there are bookshelves, there will be books." Not everyone covets bookshelves for books, however. Anne Fadiman noted that because her parents had about seven thousand books, "Whenever we moved to a new house, a carpenter would build a quarter of a mile of shelves; whenever we left, the new owners would rip them out." When Thomas Jefferson's books were sent to Washington, D.C., to help replenish, after a fire, the collection of the Library of Congress, the bookshelves—which were in fact pine boxes that could be stacked to make a bookcase—went along with lids nailed across their fronts to keep the books in place.

During the Renaissance, shelves of all kinds increasingly came to be used to display objets d'art and specialized collections of all kinds. In the early nineteenth century, James Nasmyth, the Scottish-born engineer responsible for the steam hammer, wrote of his father's retreat from his artist's studio to a space full of other things: "The walls and shelves of his workroom were crowded with a multitude of artistic and ingenious mechanical objects, nearly all of which were the production of his own hands." This tradition has continued into the present time among collectors, and it is not uncommon today to find inside a house a room encircled with shelves that hold everything from model trains to dolls, but with not a book to be seen upon them. (The household that does devote the shelves in a public area to collectibles is very often also in possession of a number of specialized directories of dealers, catalogs of wants and offerings, and books of model numbers and prices, but these are likely to be in a bedroom, where corner tables and entire corners—studies of sorts—might be given over to the working library and where the collectors pore over them on their way to sleep.)

A coffee-table book that records in striking photographs the desks of famous people, mostly writers, shows the office of Admiral William J. Crowe, Jr., who at the time of the photo was chairman of the U.S. Joint Chiefs of Staff. Along the entire wall behind his desk is a striking bookcase full of hats from the admiral's collection. There are hats, most with a military flavor, from around the world—but nary a book to be seen. (On closer inspection, the photo does in fact capture a few books—what look like a desk dictionary and a *Familiar Quotations*—but they are as inconspicuous as a palace guard's unblinking eyes when lost in the dark shadow of a ceremonial headdress pulled down over them. When those same eyes are caught in the gaze of a child, however, they are the focus of attention, as are Admiral Crowe's books, once noticed.) Similarly, the shelves behind the worktable of the illustrator David Macaulay are lined with toys, models, and artifacts of every kind but books.

Most of us do use our bookshelves primarily for books, and that is the focus of the present story; this tale will necessarily also involve the story of the book, which is both a deceptively simple and an amazingly complicated artifact. To make the terms for the various parts of the book unambiguous in this one, by the "back" of the book we should understand the part that touches the table when a book is placed with its cover up, ready to be opened and read. When a book is vertical on the shelf, the "bottom edge" is the part touching the shelf, and "top edge" is uppermost. The part facing back into the bookcase is now paradoxically but correctly termed the "fore-edge," because for a long time it was the edge placed facing forward and outward. Finally, the part of the book we now find ourselves looking at when we face a bookshelf full of books is the "spine." For centuries, the spine was shelved inward, which is among the most curious facts in the history of the lowly bookshelf and one of many that make its story intriguing.

The story of the bookshelf, as well as that of how it holds books, is one of an object that takes on meaning only in a context, only from its use. Is a horizontal board a bookshelf if no book is placed upon it? The question points to one distinguishing feature between technology and art, for the former always must be judged through the lens of utility, whereas the latter can be regarded with an eye to aesthetics alone. The most beautiful bridge that cannot carry traffic is not a technological achievement and is only arguably an aesthetic one. The most handsome bookcase that collapses under a load of books is not a bookcase but a structural failure. And just as we may wonder if a tree makes a sound when it falls out of earshot, so we may ask, Is an empty bookshelf an oxymoron?

The stories of the evolution of the book and the bookshelf truly are inseparable, and both are examples in the evolution of technology. More than literary factors, technological factors—those relating to materials, function, economy, and use—have shaped the book and the furniture upon which it rests. The evolution of the bookshelf is thus paradigmatic in the history of technology. But because technology does not exist independent of the social and cultural environments in which it is embedded and which it in turn significantly influences, the history of a technological artifact like the book or the bookshelf cannot be understood fully without also addressing its seemingly nontechnological aspects.

The description of the changing ways in which books have been made, cared for, and stored over the last two millennia provides an inherently interesting and technically simple vehicle for understanding how technology evolves. It thus provides a means for better understanding present technologies, which so catch us up in their and our own developments that we may have a difficult time observing more than just the relative changes that are occurring to us or them in the course of their everyday use. Better understanding the mechanisms of technological evolution makes us better able to understand what is happening in today's technologies and thus better able to anticipate what will happen in the years to come. Such understanding is always of value, whether we are investing in the stock market, designing and selling products, or just seeking to get a better grasp on how the world works.

CHAPTER TWO

From Scrolls to Codices

In ancient times, books did not exist as we know them today. Roman writings were turned into rolls or scrolls, mostly of papyrus, which were termed *volumina.* It is from the Latin singular *voluminum* that our English word "volume" comes. Both the width and unrolled length of a scroll varied, as do the height and "length" of a book today. On average, a scroll may have been from 9 to 11 inches across, and the total length of a volume could be in excess of 20 or 30 feet, with a given work occupying several rolls or volumes.

Greek scrolls were similar; it has been estimated that Homer's *Iliad,* for example, would have filled about a dozen rolls, and a reconstructed first- or second-century version of the complete work occupies "nearly three hundred running feet of papyrus." Had the words had spaces between them, as they do in all modern books, another 30 feet of papyrus might have been required. "It is extraordinary that so simple a device as the separation of words should never have become general until after the invention of printing," but such an observation just reinforces how accustomed we have become to practices that once were far from obvious or necessary. Wordsruntogether are foreign to our eyes, but "with a little practice, it is not so difficult to read an undivided text as might be supposed."

Scribes and scholars of antiquity have often been represented with scrolls on which the writing appears to extend across the width of the papyrus, as in some Renaissance depictions of St. Jerome in his study, but such an arrangement was more often used for tabular data, proclamations, and the like. In fact, the text of a scroll of prose typically consisted of lines parallel to the long edge of the scroll, and these lines were arranged in columns of a length and width convenient for reading, much as they are in the pages of a modern magazine.

The Latin or Greek volume was read from left to right, and when the scroll was held in the hands, the already-read portion was often rolled up in the left hand while the still-to-be-read text was unrolled from the right, not unlike the way we handle the pages of a book being read today. Sometimes the finished part of the volume was collected behind the scroll, in the way some people fold the pages of a magazine behind it, but more commonly both the read and unread text were rolled up and unrolled on the same side of the scroll. The former configuration is commonly seen today in schlock printing, the latter in the way blueprints might be unrolled at a construction site. However oriented, scrolling on computer screens takes its name from the way scrolls worked, and no matter the manner in which it was read, when a scroll was finished it would have to be rewound to be read again, very much as with a modern videotape after it is viewed.

A stick or rod was sometimes attached to the end of a scroll, with the rod often projecting some distance beyond the edge of the papyrus, thus protecting it from being crushed. When held behind pegs inserted into holes in a desk or table, such rods could be very helpful in keeping a volume open for reading or copying, freeing the hands for writing or other tasks. Stones or other objects—what today we would call paperweights—could be used to accomplish the same end. On shorter volumes, the end rods also, by their weight, could allow a scroll to be draped over the sides of a desk or table, with gravity being called upon to keep the scroll taut and opened to a place. In this case the rods themselves could also be described as paperweights.

When not being written in or read, in ancient times a volume was kept rolled up and tied with string or fastened with straps. Some more valuable rolls were fitted with a sleeve, much as a gift edition of a book might be with a slipcase today. A group of related scrolls might be placed upright in a container not unlike a modern hat box; they would look like a collection of tight spirals when the top of the box was not obscuring them. The ends of the scrolls were fitted with tags or tickets, not unlike modern price tags, which were marked to give the nec-

One way of reading ancient works was to roll up the finished text behind the scroll, as shown in this drawing made from a Pompeiian fresco.

essary information, such as descriptive words, author, and the like. A particular box might hold all the rolls containing a single work, in which case the tickets would identify the different volume numbers. Keeping scrolls in a box was convenient not only for holding them together and protecting them but also for transporting them from place to place.

Not all scrolls resided in slecves or boxes, however, and a room where such books were stored might be fitted with wall shelves subdivided into pigeonholes. An ancient library having such "bookcases" filled with scrolls lying flat on the shelves, ends out, must have looked somewhat like the stockroom of a modern wallpaper shop. The collection of scrolls making up a given work might occupy one pigeonhole fully, whereas a number of one-volume works might be kept in another, all identified by their tags or by their location in particular pigeonholes. Care needed to be taken not to pile the pigeonhole too high with scrolls, lest the bottom scrolls be crushed. (For centuries there were debates over whether Hebrew scrolls should be stored horizontally or vertically. Most synagogues keep them behind the closed doors of an ark, arranged in a vertical position so that a less holy book does not rest atop a holier one.) Such concerns also affected the shelving of early books with heavily decorated covers, which would be

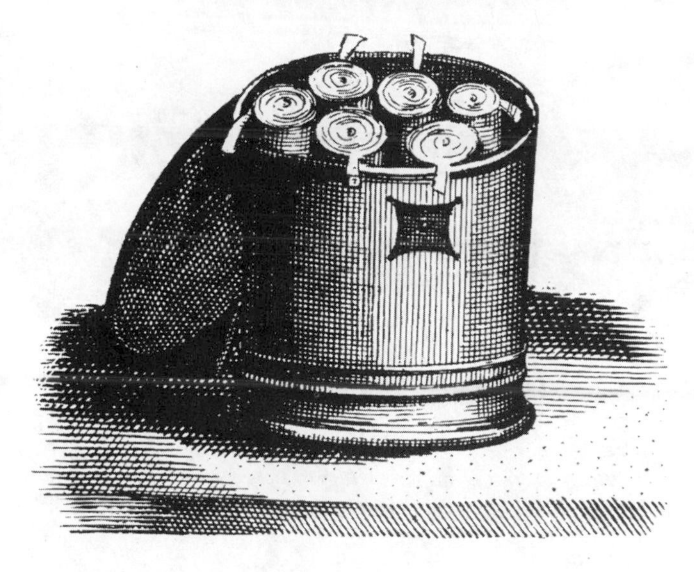

Scrolls were sometimes kept in hat box–like containers known as capsae, *as depicted in this drawing made from a fresco in the ancient city of Herculaneum.*

placed flat on shelves that did not have a great deal of vertical clearance, so that they would not be piled upon each other, or stored on narrow shelves, leaning almost vertically against the wall.

The library at Alexandria, which was founded around 300 B.C. as a repository for copies of all the books in the world, is believed to have held hundreds of thousands of scrolls at one time. Whenever a ship came into port, its scrolls were copied for the library. One story, perhaps apocryphal, has the works of Sophocles, Euripides, and Aeschylus borrowed from Athens in order to make copies for Alexandria. According to the account, when the copies were completed, it was they that were sent back to Greece, with the original scrolls being kept in Egypt. In any case, the shelving of Alexandria's vast collection of scrolls must have presented considerable problems for its librarians. The situation was no doubt more manageable for smaller, private libraries of the time.

According to Cicero, writing to his Greek friend Atticus, who had loaned him two assistants to help his own Tyrannio fix up the library

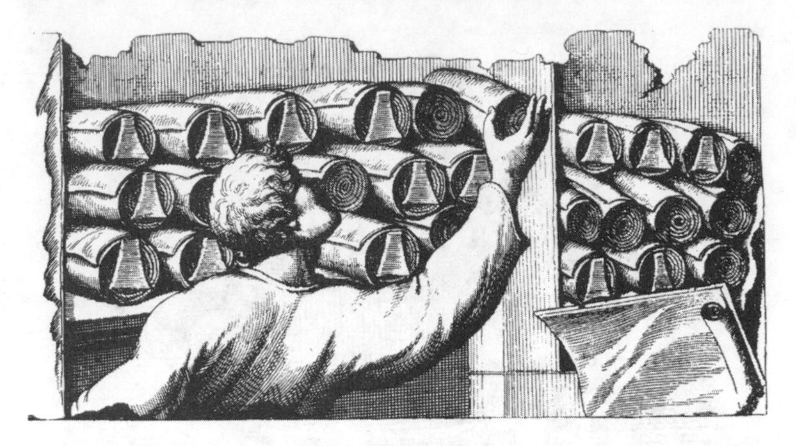

Scrolls, their contents identified by tags or tickets attached to their ends, were commonly stored on divided shelves.

by building shelves and adding tickets to the volumes, "Your men have made my library gay with their carpentry work and their titles. . . . Now that Tyrannio has arranged my books, a new spirit has been infused into my house. . . . Nothing could look neater than those shelves. . . ." The pleasure achieved through the construction and use of bookshelves would be a theme among bibliophiles for millennia, as would the penchant of some book owners to put style over substance. But not everyone was so cheerful as Cicero about how books looked on bookshelves.

Seneca the Younger, the Roman statesman and philosopher who lived a few generations after Cicero, wrote of the "evils of book-collecting":

> How can you excuse the man who buys bookcases of expensive wood, and piling into them the works of unknown, worthless authors, goes yawning amongst his thousands of volumes? He knows their titles, their bindings, but nothing else. It is in the homes of the idlest men that you find the biggest libraries—range upon range of books, ceiling high. For nowadays a library is one of the essential fittings of a home, like a bathroom. You could forgive this if it were all due to a zeal for learning. But these libraries

of the works of piety and genius are collected for mere show, to ornament the walls of the house.

How often the classics have a familiar ring to modern ears. Today we might, like Seneca, criticize the library with too perfectly matched bindings that are never cracked open or, perhaps worse, the library of books bought through an interior decorator by the yard and color, or, perhaps still worse, by the spines separated from old books and glued above "bookshelves" painted on the wall.

Ancient bookshelves were described by the fourteenth-century Benedictine monk and lover of books Richard Aungerville, also known as Richard de Bury, who became bishop of Durham. He wrote in his *Philobiblon,* which has been translated as *The Love of Books,* of the rejoicing that accompanied the restoration of Greece after a time of war. Mothers welcomed back their children to rebuilt homes, to which books had also returned:

> Their old homes were restored to their former inmates, and forthwith boards of cedar with shelves and beams of gopher wood are most skillfully planed; inscriptions of gold and ivory are designed for the several compartments, to which the volumes themselves are reverently brought and pleasantly arranged, so that no one hinders the entrance of another or injures its brother by excessive crowding.

By the early centuries of the Christian era, bookshelves had to accommodate, in addition to scrolls, a growing number of bound manuscripts, or codices, which in time would displace scrolls as the preferred format for books. The codex, named for the fact that it was covered with wood (*codex* means "tree trunk" in Latin), and which led to the term "code" in a legal context, was made by folding over flat sheets of papyrus or parchment and sewing them together into a binding. This had several distinct advantages over the scroll. Where an entire scroll might have to be unrolled to find a passage near the end, the relevant page could be turned to immediately in the codex. Also, writing in a scroll was normally on one side only, whereas the codex lent itself to the use of both sides of the leaf.

The codex evolved from the tablets made of wood or ivory that in

Wooden tablets, or table books, were precursors to codices and the modern book.

classical times were hinged together to form what might be described as a portable writing surface. Tax collectors and others who needed to make notations while standing or while sitting on a horse would have found rolls unmanageable. Not only did scrolls have to be kept from returning to the natural, rolled-up position, but they also needed a hard surface backing them. In comparison, the handheld tablet was ideally suited for note-taking. It could be immediately opened to the desired place, and it presented its own hard surface on which to write. The writing was often done with a stylus on a prepared or impressionable surface. Rather than needing a third limb to hold an inkpot, everything could be done easily with two hands. When the task was done, the tablet could be tied or clasped shut to protect its contents, and carried securely. Some tablets had hollowed-out "pages" filled with wax, so that after a day's notes were transcribed to a more permanent record, as to a scroll, the impressions in the wax could be smoothed out with the flat end of the same stylus that had been used to make them, and the fresh tablet book was ready for another day of note-taking.

The earliest codices, which took their form from the informal tablet book, apparently date from the early part of the Christian era (about the second century), and it has been speculated that the codex form might have been first adopted when the Christian Bible began to be copied on papyrus and circulated in book as opposed to roll form to distinguish it from the scrolled texts of Judaism and paganism. With its clear advantages over a scroll, "by the beginning of the fourth century, the codex became the predominant medium for both Christian and non-Christian literature, and the use of the roll sharply diminished."

There was, however, a transitional period when book owners had both scrolls and codices to contend with in arranging their libraries, and this clash of forms may have been what drove the widespread adoption of the closed cabinet. For all their ability to dress up a room and impress the neighbors, open shelves came in time to be viewed as unsightly and so were enclosed behind doors. This may have been done for several reasons, all related to problems that their owners could have experienced or anticipated with regard to keeping books exposed, no matter how they cheered up a room or made it fashionable. Indeed, the fact that Cicero found his remodeled library so remarkable suggests that he had not thought it previously to be as attractive as he wished. With Atticus's library as an appealing model, therefore, Cicero had his own spruced up.

Some other book owners might have found unattractive the exposed ends of rolls labeled in their cases (bindings) or tagged with tickets, and so they added doors to hide the utilitarian objects the cabinet held. Still others might have worried about moisture accumulating, or dust collecting on the rolls, or vermin crawling into them. And still others may have had trouble with thievery or unauthorized borrowing of their scrolls, which may only have been discovered when a needed volume was nowhere to be found. When codices as well as scrolls had to be accommodated, the two dissimilar forms would have presented a dissonance on the open bookshelf, further encouraging the addition of doors.

The closed cabinet was known in Latin as an *armarium*, and the word "occurs commonly in Cicero, and other writers of the best period, for a piece of furniture in which valuables of all kinds, and household gear, were stowed away; and Vitruvius uses it for a bookcase." In time the word came to designate a form of cupboard, wardrobe, or closet, and came to be rendered into English in many variants, including "almery" and "armoire." Another word, "press," came to be used increasingly to designate a bookcase, especially one comprising several sections, with the etymological trail from the Old French via the Middle English *presse*, a thirteenth-century word referring to a crowd or crowded condition, suggesting that the problem of finding space for books is not a new one. We still refer to presses for china, linen, and the like, but American librarians tend to consider the word a Britishism.

The doored book press—the word "press" not to be confused

with its meaning associated with printing—could be closed and secured; it could hold the scrolls and, perhaps more importantly, the newer and rarer codices that in many cases were also more valuable to their owners. An alternative and more portable means of securing scrolls and codices was a book chest, akin to a trunk.

Several literary traditions also coexisted with scrolls and codices, and in some libraries "the Roman classics were arranged on one side of the room, and the Church Fathers on the other, much as the Greek and Latin collections were kept separate in the Roman temple libraries." As there were different literary traditions, so there were also different materials used in the creation of written documents. The media have been diverse indeed and have included:

> small stones; large stone faces afforded by walls of caves and natural cliffs; clay tablets and cylinders; bricks and tiles; bark, wood, palm-leaves and papyrus; linen; wax tablets; metal, ivory and bone, together with leather, parchment and raw-hide produced from animal skin, and in more recent times, paper.

For a long time papyrus was the medium of choice. The word is believed to be of Egyptian origin, as is the plant. The Greeks referred to papyrus as *byblos,* after Byblus, the Phoenician city that was a center of papyrus exportation. Hence we have the Greek word for book, *biblion,* which in turn gave us the English word "bible," "The Book."

Papyrus was made from the solid-stemmed marsh plants that grew along the Nile River. They were split open and beaten into flat sheets that could be joined end to end to form rolls of any length. At first, papyrus was also used in codices, and it continued to be utilized in Egypt "well into the tenth and eleventh centuries." But this lightweight material had its disadvantages as well as its virtues:

> Papyrus was cheap and abundant in the Greek world, and it solved the problem admirably for several centuries. It suffered from two handicaps, however. Except in the dry climate of Egypt, it was very perishable. Damp ruined it, and in Rome, and still more in Gaul, texts had to be constantly re-copied to preserve them. The Emperor Tacitus, for example, wishing that the works of his imagined ancestor, the historian of the same name, should be in all the Roman libraries, felt it necessary to order ten copies

> to be made by his official copyists every year and delivered to the libraries. And Martial in his epigrams frequently reminds his readers that a shower of rain would damage his books, while to use them (as some apparently did) for wrapping fried fish, produced immediate disintegration. The other difficulty was that papyrus was weakened by even a single fold. It was really satisfactory therefore only for the roll form of book.

There was clearly plenty of reason to look for an alternative to clay, stone, and papyrus, but, as is often the case in the development of technology, it was extra-technological events that provided the impetus for change. According to Pliny's *Natural History*, King Eumenes II, the second-century B.C. ruler of the Greek kingdom of Pergamum, tried to import papyrus in order to establish a library that would rival Alexandria's, but Ptolemy Philadelphus did not allow exportation of the material from Egypt. Eumenes, not to be deterred, ordered that sheepskins be processed into smooth thin sheets that could be used in place of papyrus. According to Pliny, the material was called *charta pergamena*, after its place of origin, which in turn led to the word "parchment." In fact, however, parchment was not invented in Eumenes's time, but only readopted then; it had earlier been known as *membrana*. In modern usage, parchment has come generally to refer to any writing medium "made from sheep, goat, or other animal skin."

Vellum was another alternative to papyrus. Although vellum and parchment are often confused in usage, they are, strictly speaking, distinct materials. Vellum, etymologically related to "veal," is made from calfskin, though the Latin term relates to the hide of sheep and other animals as well. Indeed, "the skins of almost all the well-known domestic animals, and even of fishes, have been used for the purpose of making a material for writing upon," with stillborn lambs and calves having provided "some of the finest and thinnest" material. In the final analysis, vellum and parchment proved to be more durable than papyrus. Unfortunately, the animal-derived material did not come easily, for "one sheep yields no more than a single sheet (two leaves) for a folio book." Thus, "a very large flock of sheep" might have to be slaughtered to obtain the parchment needed for a single codex.

The ability of animal skins to hold up under stitching meant that pieces of them could be cut to standard sizes and sewn together to form rolls or "books" made of "stitched sheets folded in concertina fashion" or codices. Although it has been argued that the "rectangular" shape of animals "dictated the shape of the books—a convention we still have today," in fact even the earliest codices were rectangular, a shape derived from folding and stitching together sheets of the vegetable material papyrus, whose own shape derived not from the plant's but from the method of manufacture for use in rolls. A rectangle can be doubled over in two ways, however, making the folded shape taller than wide or wider than tall, with the writing arranged in what in the computer age is referred to as the "portrait" or "landscape" mode, respectively. Since the codex developed from the scroll, it would seem to have been more natural to produce early codices in the landscape mode, which would have had the columns of words most nearly imitating how they were arranged on a scroll. However, the need to bind the folded sheets along one edge would have argued for the portrait mode, because it located the stitching along the greater dimension, thus providing maximum strength to the binding.

Virtually all early books were thus made in the portrait mode that still dominates book manufacturing. Two unusually shaped books produced in the late twentieth century demonstrate in an exaggerated way the different effects of binding on the long and short side. Judith Dupré's *Skyscraper* is, like the building form itself, narrow and tall, with a page size of about 7½ inches by 18 inches. As is appropriate for its subject, the book is bound along the larger dimension, and this makes for a binding that gives great support to the pages, no matter what position the book is in. The book is stiff on the bookshelf and in the reader's hands, giving the look and feel of a well-constructed tall building. Dupré's later book, *Bridges,* as is fitting for its subject, is long and squat, however, and its 18-by-7½-inch pages are bound along one of the shorter sides. This is a book that must be handled with care.

No matter what the origin of the shape of the material on which an ancient text was written, early codices usually were bound literally between wooden boards, often beveled, cambered, or chamfered at their edges and having grooves and channels gouged out to provide a means of anchoring the book proper to its covers. Though clearly related to the tablet book form, the boards of codices would likely have been larger and thicker than those used in the hand-sized tablet notebooks. Codices would thus have been much heavier and some-

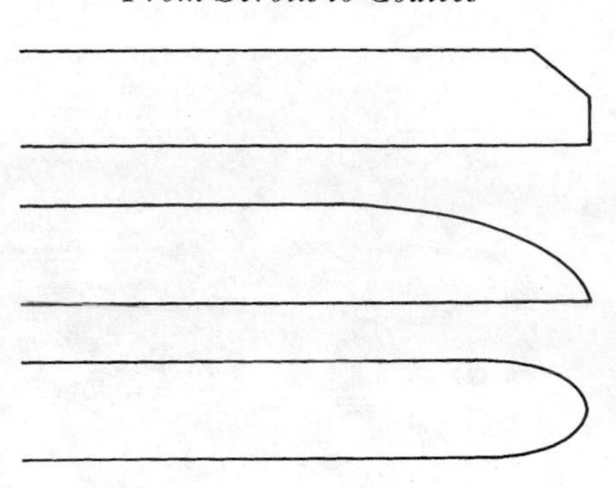

The edges of wooden book covers were finished by being beveled, cambered, and chamfered (top to bottom).

what unwieldy to use, but they would definitely not have been so inconvenient to read as scrolls.

An ancient codex was often stored on a table or shelf, frequently one sloped akin to a writing surface. On such an inclined surface, the book's front cover could be displayed as a work of art as well as a distinctive identifying case for the inscribed leaves inside. The binding may have been of parchment, leather, fabric, or other covering over the boards. Books especially valued were covered with "treasure bindings"—bindings heavily jeweled or otherwise lavishly decorated—usually because the contents were thought to warrant it for religious or ceremonial reasons.

Seldom was any title or author identified explicitly on a book's cover. In fact, codices did not have titles as we know them; the books were referred to, perhaps, by the first words on an early page. The work that we know today as *De rerum natura,* for example, may well have been identified by Lucretius's opening words, *Aenadum genetrix.* Neither were there page numbers; passages were located for reference by key words in the text.

When a scholar did find a passage he wanted to study or copy, whether in a scroll or codex, it was convenient to have some means of holding the place to free the hands to hold writing implements. Scrolls might have been held in many clever devices, such as the slotted lectern that St. Jerome is shown using in the manuscript *Les Miracles de Nostre Dame,* written at The Hague in 1456. Though the

A scroll is held open by a slotted lectern in this fifteenth-century depiction of the scholar St. Jerome (c. 347–420).

technology may or may not be ancient—the Renaissance artist possibly took some liberties in depicting artifacts without regard to their place in history—the drawing does illustrate the principle of furniture being adapted to the needs of the book. In some cases, the weight of the papyrus alone could be used to hold it in place for the reader or scribe. If necessary, weights at the end of a string could be draped over a scroll to flatten it further and help keep it in place. This has been illustrated in a drawing of Jean Mielot, a fifteenth-century scribe who was secretary to Philip the Good, duke of Burgundy—though, unless Mielot is preparing some ceremonial document, the depiction seems curiously anachronistic, since he appears to be copying onto a scroll from a codex. (The codex Mielot is working on is also shown being held open to the proper place by a weighted string, a useful device still employed in rare-book libraries today.)

Mielot is clearly shown working at a pup-tent-shaped desk over which the scroll is draped and stays in place. Although the slope of the desk seems to be extreme in this case, being almost vertical, it appears to have been more desirable to have the working surface as upright as possible, just as an artist often prefers to work on a painting held almost vertically on an easel. Indeed, ancient and medieval calligraphy may be viewed as an artistic rendering of language, with the calligrapher functioning more as a draftsman than a writer. A scribe, in fact, did not even have to be literate to copy a work verbatim.

The use of an inclined desk surface, albeit more moderate in slope, was continued through the nineteenth century and can still be found in some contemporary desk designs, as well as in typewriter keyboards. Today, many computer keyboards have retractable feet that can be locked in an extended position to give the rows of keys a tiered effect, suggestive of that on old manual and electric typewriters. Since the keyboard of my laptop computer is not tiered, I achieve the sloped effect by propping up the back of the computer on a book. This brings the upper keys more comfortably into reach. Alternatively, I can lean over the keyboard to get the same effect, but that position becomes tiring after a while. Even when I am not typing at a keyboard, I find a sloped surface desirable. When I was revising an early draft of this book, I sat in an easy chair with my right leg crossed over my left or propped up on a stool to provide an inclined surface on which to rest the manuscript when writing upon it. No matter what its form, the sloped desk or shelf has played a significant role in the development of the bookshelf, as we shall see.

With the exception of the continued use of the scroll in the practice of religion and for legal purposes in a country like Britain, where there remains a Master of the Rolls, the codex in time did drive out the scroll—general texts being copied from rolls into codex form as early as the fourth century—and thus shelves and armaria came typically to contain only volumes more recognizable today as books. With the increased number of books with which libraries of all kinds had to deal—and collections always do seem to expand—furniture to hold

The fifteenth-century French scribe Jean Mielot is shown writing on a scroll. It was more common for scrolls to be composed in columns, as in the open codex on the lectern-shelf to the right. In fact, it was the arrangement of writing on scrolls that dictated the columnar form of the text in codices and, later, early printed books.

the books multiplied and grew larger. Armaria generally retained their form of being essentially what today we might call cupboards or wardrobes, and increasingly in the Middle Ages they were kept locked or otherwise secured.

Security was necessary, of course, because every book was produced by hand. Each letter, word, sentence, paragraph, page—each entire volume—was laboriously executed by a scribe, either from another manuscript or from the dictation of a lead scribe who presided over a stable of book producers, much as a master may have supervised the galley slaves rowing an ancient trireme. The dictation system lives on today in the classroom, where students copy down in their notebooks the words of the teacher, sometimes verbatim—especially when the teacher writes on the blackboard and mutters something about the material possibly being on a test—but more often than not in some telegraphic form. The manuscript system also survives at news conferences, where the words of a public figure are copied down by the journalists in attendance, who are jocularly known as scribes. Even if the image of the newsmaker is being transmitted electronically to and from a satellite and the reporters are using tape recorders or laptop computers, the news conference still operates on the medieval model of book production.

Today, of course, whether the information comes from a press secretary standing at a lectern or from an astronaut floating behind a microphone in space, it can be instantaneously transmitted via radio, television, and the Internet around the world. What a reporter with a laptop composes on the spot can also be downloaded to the newspaper's computer, edited on a screen, and set in type automatically, without a single word appearing on a single piece of paper until the morning edition comes off the printing press. Such modern journalistic practices may seem to be a long way from a scribe copying or a Gutenberg setting lines of type from a manuscript bible, but there is a certain sameness to it all. That is what makes the history of technology interesting and relevant: it not only teaches us about the way things used to be done; it also gives us perspective on how things are done today—and how they most likely will be done in the future.

CHAPTER THREE

Chests, Cloisters, and Carrels

The disciplined life of a monastery, not to mention the religious commitment of its residents to prayer and to study of the Scriptures, made it a natural venue for the production and preservation of manuscripts. But that is not to say that in the early Middle Ages monasteries were full of books. A monastery's entire holdings might be counted in the tens of codices. A library consisting of hundreds of volumes would be a great one indeed. We can imagine that it was a significant event whenever a new manuscript was added, perhaps by being copied from a book borrowed from another monastery. A codex could also be acquired in trade for a book duplicated from one of a monastery's own books, or as a gift or bequest.

The very remoteness of some monasteries alone would provide a degree of protection for its books, but even within the confines of the cloister there were highly regulated procedures for keeping track of the collection. It appears to have been common for a monastic order to have a librarian—sometimes called a "precentor," a word that also designated the leader of singing, who would have had to keep track of hymnals, psalters, and the like. The librarian/precentor was thus responsible for knowing where the order's books were at any given time. From the early Middle Ages, some orders had a custom similar to that of the Benedictines, in which the members of each chapter assembled at a predetermined time to return books that had been issued to them during the previous year and to borrow a new book for the coming year. According to an eleventh-century description of the "general monastic practice" of English Benedictines,

> On the Monday after the first Sunday in Lent, before brethren come into the Chapter House, the librarian shall have had a carpet laid down, and all the books got together upon it, except those which a year previously had been assigned for reading. These brethren are to bring with them, when they come into the Chapter House, each his book in his hand. . . .
>
> Then the librarian shall read a statement as to the manner in which brethren have had books during the past year. As each brother hears his name pronounced he is to give back the book which had been entrusted to him for reading; and he whose conscience accuses him of not having read the book through which he had received, is to fall on his face, confess his fault, and entreat forgiveness.
>
> The librarian shall then make a fresh distribution of books, namely, a different volume to each brother for his reading.

The present custom of university libraries to reconcile at the end of each academic year charges to faculty—who like medieval monks are often allowed to keep books for extended periods of time—traces back to the Benedictine practice. The possibility that a book might not be read—or copied—in a year suggests the low level of literacy or scholarly curiosity that may have existed in the monasteries, as it may even in some institutions of learning today. Nevertheless, the fact that a carpet was laid out attests to the care that was taken with the books, whether they were read or not. A similarly watchful custom was followed by the Augustinians:

> The Librarian, who is also called the Precentor, is to take charge of the books of the church; all which he ought to keep and to know under their separate titles; and he should frequently examine them carefully to prevent any damage or injury from insects or decay. He ought also, at the beginning of Lent, in each year, to shew them to the convent in Chapter. . . . He ought also to hand to the brethren the books which they see occasion to use, and to enter on his roll the titles of the books, and the names of those

> who receive them. These, when required, are bound to give surety for the volumes they receive; nor may they lend them to others, whether known or unknown, without having first obtained permission from the Librarian. Nor ought the Librarian himself to lend books unless he receive a pledge of equal value; and then he ought to enter on his roll the name of the borrower, the title of the book lent, and the pledge taken. The larger and more valuable books he ought not to lend to anyone, known or unknown, without permission of the Prelate.

Where were "the larger and more valuable books" kept in the Middle Ages, when "books were rare, and so was honesty"? Not only were libraries small but also the individual volumes were not easily replaceable, and so a common way to keep books was to lock them up in an armarium or a chest resembling a foot locker when they were not in use or checked out to a responsible individual. The chest is believed to have been employed contemporaneously with the armarium, but used for smaller collections and those that had to be transported. Three book chests in Hereford Cathedral, in western England near the Welsh border, survived into the nineteenth century, when they were recognized as such. One of them, which has been dated from around 1360, measures about 6 feet long, 21 inches high, and 21 inches wide. The chest is elaborately carved and has stout feet at each corner. The top is secured by three different locks and thus required three different keys to open it. (The way locks were fashioned at the time would have made it unlikely that a single key would open any more than the single lock that it was produced to fit and operate.)

Chests of this kind, even when full of manuscript books, would not deter the most determined of looters, of course, for the entire chest could be lifted up and carried off the premises. Furthermore, the wood could be shattered easily with a hefty ax. The purpose of the chest was not so much to protect the books from wholesale thieves—for those were to be kept off the monastery grounds—as to secure the books from surreptitious borrowers who might not remember or wish to remember that they had, for whatever good or questionable reason, removed a particular volume. The purpose of the locks was to keep the lid from being opened by unauthorized persons. More important, however, if the three keys were held by three different monks, no one or even two of them could open the chest without the other one or two knowing. One of those keeping a key would naturally have been the librarian, who would thus know at all times who was removing a book from the chest, and so could record the action. The elaborate lengths to which medieval monks may have gone to protect—or suppress—books in their libraries have been vividly imagined by Umberto Eco in his novel *The Name of the Rose.*

Another medieval book chest, which survived at Hereford by being used as a storage place for everything but books, can be dated from the fourteenth century, judging by its ironwork. The chest is made of poplar, a very light wood, but in its modern state it has been refitted with a heavy oak lid. The original construction would have made the chest relatively easy to transport, and the iron straps and corner reinforcements would have toughened it against damage in transit. The straps also would have helped distribute the load of the chest's heavy stock of books, which would have had to be carried from place to place as the bishop moved—with the necessary books—among his various residencies. The chest is almost 4 feet long, 18 inches high, and 20 inches wide. Thus these two Hereford chests share most nearly their width dimension. This may suggest something about how books were arranged in the chests. The smaller chest, which does not have carving that might suffer distress in its transport, was also fitted with end rings through which a carrying pole could be inserted. Another chest at Hereford has a top formed out of a tree-trunk segment, and thus has a rounded, and bark-covered, rather than a flat top.

It was not only traveling bishops who used book chests. Royalty also used them, as did the inhabitants of monasteries. An illustration from a twelfth-century illuminated manuscript shows Simon, the abbot of St. Albans, seated before a book chest, reading from a book—or possibly showing someone reading over his shoulder a point for which the book was consulted in the first place. The lid of the chest is open, further suggesting that the book had been removed from the chest, which clearly holds other volumes. One of the locks of the chest is shown near the right end of it, which suggests that there is also a lock near the opposite end and possibly a third in the middle. (Two hasps in fact are clearly visible on the partially obscured lid.) It is likely that in order for the chest to be opened at all, at least one other individual had to be present with a key—unless the abbot had both keys. There is what appears to be a miter resting on the open lid; the

In medieval times, books were often kept in chests, such as this one before which Simon, twelfth-century abbot of St. Albans, is shown reading.

abbot, who could have worn such a headdress, may have had to lean into the chest to find the book he had retrieved and did not want his hat to fall or be knocked off in the process. The illustration provides a view of a bit of the inside of the chest, and a book in it appears to be standing on its spine, i.e., with its fore-edge turned upward. This may be because the abbot had been rummaging around to find the book he was looking for, but it is also possible that the books were stored in the chest in that position; the fore-edge, not the spine, was more likely to have carried some identification of the contents.

Simon appears to be supporting the book he is reading on the front edge of the chest, which is at a convenient height. The chest appears to have been deliberately raised to such a level by being set upon a frame of some kind. Medieval book chests generally, or at least those that have survived in fact or in illustration, either have feet or are raised on frames, for at least two reasons. First, raising the chest at least an inch or so off the floor enables it to be picked up more easily for moving and transporting. Second, keeping the bottom of the chest off the floor would protect its wood, and by extension its contents, from damp and water damage. The dampness of medieval stone abbeys was of some concern; in a record of the customs of the Augustinians, the librarian was cautioned that book alcoves formed in a stone wall should "be lined inside with wood, that the damp of the walls may not moisten or stain the books." (Books that were kept in such open alcoves were the more common ones, like psalters, which the monks used regularly during services.)

The locked chest would naturally hold books of considerable value and those books should not have been taken away from the open chest—and the reader should not even have turned his back to the chest—lest some unauthorized person be tempted to borrow a book and fail to return it before the chest was closed again. Thus, to read facing the books in the chest—as Simon is doing—was a matter of responsibility and security. Furthermore, even if the chest were in the corner of what we would consider a library, it cannot be assumed that there would be a nearby table or desk on which to lay and read the book. In fact, medieval readers most likely would be uncomfortable reading a book in a flat position, because the way they most often encountered books set out to be read was propped up on another book or on a slanted surface not unlike a modern lectern or music stand. Indeed, the word "lectern" comes from the Latin verb *legere,* "to read," and even a modern lectern has a sloped surface to hold books or notes. A podium, on the other hand, though often confused with a lectern and now sometimes used as a synonym for it, is something to stand at or upon, as suggested by the word's relationship to the word "pew," a piece of church furniture that played an important role in the evolution of the bookcase.

One thing Simon may have found wanting at his chest of books was a good source of light. By facing the wall as he did, he tended to put the book in shadow, though he does appear to be holding it away from him so as to catch as much light as possible on its pages. Long before there was electricity, there were candles and oil lamps, of course, but readers complained of their effects the way we grouse

Books in general use in monasteries were sometimes kept in a wood-lined recess in the wall of the cloister near the door leading to the church. This example dating from the late twelfth century is from the Cistercian Abbey of Fossa Nuova. The wooden shelf and doors are believed to have been removed to be used elsewhere when this armarium commune *was no longer adequate for the storage of books.*

today of second-hand smoke and noxious fumes in hermetically sealed new buildings. Thus, for a long time the preferred way of reading a book was in broad daylight. For the scholar lucky enough to have a study space situated beside a properly oriented window in a normally sunny climate, there must have been no greater pleasure than to sit at that window and read, or perhaps on a particularly nice day to take a book into the garden and sit beside some colorful and fragrant flowers. For medieval monks confined to the cloister of a monastery, with its regimen and restrictions, there may have been many such pleasures associated with the book, though those wishing some privacy may also have been distracted by them.

In time, individual monks, like many a scholar today, came to desire a private enclosed area in which to study. Such an enclosure, known as a carrel, consisted of an individual cubicle, often no larger than a broom closet, that was highly coveted because it was indeed a room of one's own. The first recorded reference to monastic carrels that has survived appears in connection with the Augustinian Order and dates from the year 1232. Carrels have been variously described as "curious wooden contrivances" and "tiny studies, about the size of a sentry box." In spite of the diminutive nature of the carrel, it was to serve as the model for the private study of the Renaissance.

How the carrel developed is an excellent example of how technology evolves within available means, and changes to deal with problems as they are encountered. The architecture of monastic cloisters and later Gothic cathedrals is familiar; often among the most photographed features are the long series of stone pillars that line the outer edge of the covered walkway open to the air and looking out on a yard or garden. The inside of the walkway is usually defined by a blank wall, behind which may be the chapel, church, or cathedral proper. There are no windows on this wall because the width of the walkway would have diminished the light reaching them, and anyway there was a source of light into the church in the windows higher up in the clerestory. By not having windows close to the floor level, the worshippers were not so easily distracted during religious services by any goings-on outside.

Where they existed, benches in the alcoves formed by the pillars separating the walkway from the yard provided the preferred place to sit and read, for it was here that the best reading light could be found. Especially in monasteries where there was no separate "scriptorium," or room devoted to writing, the better-lighted spaces recessed between the columns and pillars of cloisters became prized locations claimed by the more senior or politically astute monks, as such spaces gave a most desirable place to engage in reading, writing, or copying. This carrel, as the space came to be known, was a place to work in silence and relative solitude and thus in greater concentration on the task at hand. (Sometimes, of course, such conditions are more conducive to napping than studying.)

Western monastery carrels must have been comfortable, if not sleep-inducing, nooks in temperate regions, but they could be bracing

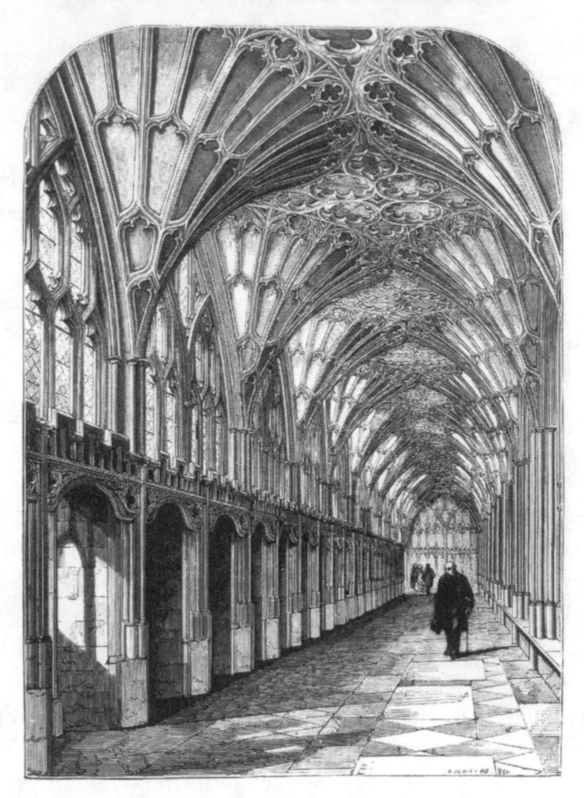

The cloister walk at Gloucester Cathedral has a long row of recesses that would have made excellent study carrels.

indeed in colder climates. There were, in fact, "plenty of complaints from scribes of the hardships of writing in the cold of northern winters." Furthermore, because they were also opened to the cloister walk or arcade, there was the distraction of human traffic. To address these disadvantages of the primitive carrel space, it came to be made warmer and separated from the traffic by wainscoting carried around to enclose the carrel. The wooden partition was fitted with a gate or door, and the space open to the courtyard came to be fitted with a glazed window. Thus, both distractions and the elements could be kept at bay. According to Burnett Streeter, a former canon of Hereford Cathedral and chronicler of medieval chained libraries, "It was in the well-lighted carrel in the cloister—not in dark 'cells' as is popularly supposed—that the monk read, copied and painted the beautiful illuminated [manuscripts] which we so much admire."

Carrels that existed at the Cistercian House at Clairvaux, France, were described in the early sixteenth century as places "where the monks write and study." But by the early eighteenth century, in this monastery at least, they were no longer a place of silent meditation and study:

> From the great cloister you proceed into the cloister of conversation, so called because the brethren are allowed to converse there. In this cloister there are 12 or 15 little cells, all of a row, where the brethren formerly used to write books; for this reason they are still called at the present day the writing-rooms.

As with all technologies, there appear to have been abuses in the use of medieval carrels, as their occupants kept books behind closed doors and thus less readily available for use by others, a clear transgression of library etiquette. Nevertheless, the benefits of the closed spaces for serious work constituted such a clear advantage that the system of building and using them grew even as egregious transgressions argued against their being sanctioned.

Where there was a limited number of carrels (as there eternally seems to be), they were claimed by or assigned first to those who were most in need of the space—or those who appropriated it by virtue of seniority. According to Abbot Ware, who wrote in the mid- to late thirteenth century, it was only when novices attained a certain degree of proficiency that they could sit in the cloister and "be allowed to glance at books taken out of the presses (*armaria*) belonging to the older monks. But they must not be permitted as yet to write or to have carrells," in part because they were "places where private, and possibly illicit, property could be kept." Indeed, "the appropriation of carrels by individual monks who kept them locked was the cause of many complaints, and there were frequent episcopal orders that they should be regularly inspected three or four times a year."

To this day, there are carrels in libraries of all kinds, and in some research libraries they are little different than they were in the Middle Ages. At Duke University, where I have for many years been assigned a closed study and work space, the modern Gothic addition to Duke's

Perkins Library has highly desirable carrels located throughout its stacks. (The original library also has carrels, but none of them has a window now because the windows were covered in the course of constructing the addition.) The size of the closed and lockable carrels varies with their location within the bookstacks. Those that are against inside walls of the stacks, and hence are windowless, tend to be about the size of a clothes closet, and thus are barely large enough to accommodate a small desk and chair. Privacy is provided by 8-foot-tall oak paneling, but the compartments are open at the top.

Such carrels are possible in a modern library, of course, because of the existence of artificial lighting. When I was first assigned a carrel, it was an inside one, and it was about as far from any window as it possibly could be. There were many bookshelves between my carrel and the building's outside walls, so it was extremely dark when the lights were out because of, say, a power outage, or when the library was closed. On several occasions I arrived at the library earlier than it opened and walked in with the staff. Since the lights in the stacks had not yet been turned on, I had to grope my way to the carrel and fit the key into the lock by touch and habit. Only when I was inside could I turn on the small lamp and have enough light to read and write.

The largest and most desirable carrels in Perkins Library are located against outside walls, with dimensions determined, just as they were in medieval times, mostly by the spacing of the pillars between the windows or the mullions separating a window into vertical parts. For years these carrels were the object of my envy, but all I could do was ask the keeper of carrels to add my name to the long waiting list. In time I did receive a new assignment, to one of the window carrels along the northwest face of the building. It seemed no larger than my previous one, but it had more privacy, because its paneling extended all the way to the ceiling, and it was lighted by a tall window and so could be worked in without artificial light on all but the dreariest of days. (When the late-afternoon sun streamed through the window, however, the excessive light and heat made the carrel as difficult to work in as it was when totally dark.)

The system of checking out books to the carrel was a pleasure. I could roam throughout the stacks of this wonderful library—which then held four million volumes—and bring any book back to my carrel. The book was checked out to the carrel by my filling out a charge slip and also a tall and narrow green card, inserting them into the book with the green card sticking out like a bookmark and leaving the book on the corner of the desk nearest the door. The door had a small window, which was not to be covered (but which often was by the postcards, posters, and placards by which carrel occupants seemed to attempt to establish their individuality, announce their dissertation topic, and guard their privacy). Each carrel was checked daily through this window, and if the attendant saw that there was a book with a green card projecting from it, she used a master key to enter the carrel and process the book, leaving it for the occupant's use.

To return a book, the green card was simply turned upside-down and the book placed in the same corner of the desk. The book would be picked up by the attendant, the circulation record was adjusted to reflect that the book was no longer in the carrel, and the book was returned to the shelves. One could hardly have expected a more ideal working arrangement, but I did find one at the National Humanities Center, where I spent a year working on a book on the history of the pencil. Here, in a building laid out with long lines of studies, as its room-sized carrels are called, there was no library as such, but there was a librarian and two staff members who obtained from local research institutions and interlibrary loan any books the fellows of the center wanted, and left them in piles to be picked up off the commons area, where the books were also returned. (While I understand this system is still in place at the center, Duke's carrel charge is unfortunately no longer in effect, presumably because of staffing costs, and now I must carry books down to the circulation desk, where I can check them out as if I were taking them home. Whether I take them back to my carrel is of no concern to the circulation department.)

As my use of a carrel intensified, and the accumulation of books I kept checked out in it grew—filling the single bookshelf above the small desk and spilling over onto the windowsill and floor—I dreamed of a still larger carrel on the top floor of the library, where the neo-Gothic windows reached dimensions of about 7 feet wide and 14 feet tall at their peak. In time I did get assigned to one of these carrels, on the northeast side of the building. There is only a hint of direct sunlight in the earliest morning hours, but the vastness of the window lets in more than enough diffused natural light. Only on very rare occasions, or in the evening hours, have I had to use the single small lamp that illuminates the ample desk with its more spacious bookshelf. (There is a light fixture in the high-peaked ceiling above the carrel, but the relatively inaccessible bulb has been blown out more often than not.)

That library carrels continued at the end of the second millennium to be arranged and constructed pretty much as they were in the Middle Ages speaks to the sensibleness of their design, given the constraints under which they have existed and continue to exist. There were, however, some arrangements used in medieval monasteries that were to outgrow their space, and these had to do with how books were stored in the vicinity of the carrels so prized by the monks.

One of the logical places to keep armaria or chests of books not assigned or spirited away to carrels was against the inside wall of an enclosed arcade or cloister walk. In this position, the books would have been convenient to the workers in the carrels. Even if a librarian's key was required to get at the often heavy volumes, the books would not have had to be transported very far from storage to workspace. (French workers at the newly opened National Library in Paris—once hopefully thought of as "the first library of the third millennium"—went on strike in 1998 in part because they had to carry heavy volumes that were too large to be delivered by the computerized system, which had broken down anyway, "through heavy doors and long corridors" to reading rooms some distance from the shelving areas in L-shaped towers that resemble open books.) According to Streeter, by about the year 1300 it was established practice along "the row of carrels" in monasteries that "books, taken from adjacent almeries, could be read in a good light in a semi-public place, under the supervision of the person or persons responsible for the safety of the books."

In addition to being convenient to the carrels, book storage along the wall of the arcade took advantage of space not otherwise very usefully employed. Before the carrels were fully established, the cloister was no doubt a place of quiet contemplation and, when allowed, conversation. The stone benches built along the wall or into the alcoves and recesses formed in it would have been a convenient place to sit and think or converse. These back walls were generally undecorated and certainly did not compete for a monk's attention with the blue sky or green grass onto which the cloister looked.

When carrels began to obstruct the view to the open air, and when the presence of contemplators or conversationalists started to distract—if not annoy—those working in the carrels, the suggestion to use the space against the wall for book storage would likely have met with little or no opposition and perhaps even a little encouragement, especially where windows had come to be installed between the outside pilasters, thus protecting the cloister walk from the elements. Even with carrels in place, there must have been enough light coming in—through any gates that might have been installed to keep the carrel's contents under lock and key, or over the tops of their woodwork when solid doors were used—to allow those looking for a book to find it. The arrangement at Durham Cathedral was recorded in *The Rites of Durham,* "that curious book" that is "an account by an eye-witness of the 'Monastical Church' of Durham before the suppression," as follows:

> In the north syde of the Cloister, from the corner over against the Church dour to the corner over againste the Dorter dour, was all fynely glased from the hight to the sole within a litle of the grownd into the Cloister garth. And in every wyndowe iij Pewes or Carrells, where every one of the old Monks had his carrell, severall by himselfe, that, when they had dyned, they dyd resorte to that place of Cloister, and there studyed upon there books, every one in his carrell, all the after nonne, unto evensong tyme. This was there exercise every daie.
>
> All the pewes or carrells was all fynely wainscotted and verie close, all but the forepart, which had carved wourke that gave light in at ther carrell doures of wainscott. And in every carrell was a deske to lye there bookes on. And the carrells was no greater then from one stanchell of the wyndowe to another.
>
> And over against the carrells against the church wall did stande certaine great almeries of waynscott all full of bookes, wherein did lye as well the old auncyent written Doctors of the Church as other prophane authors with dyverse other holie mens wourks, so that every one dyd studye what Doctor pleased them best, havinge the Librarie at all tymes to goe studie in besydes there carrells.

In other words, at least at Durham, the openings to the yard on one whole side of the cloister were glazed almost from top to bottom. Into each window were fitted three separate carrels, to which the monks retreated each day to study. The small carrels were solidly paneled, but their doors had openwork through which light—and the glance of others in the cloister, including the librarian—could pass. Each carrel was no wider than the space between vertical divisions of the windows. Opposite the carrels, against the windowless wall of the church, were armaria full of books. These were apparently unlocked, and the books in them were readily available for study.

All of the light that came into the cloister space between the carrels and the bookcases would have had to come in from a single side—the glazed south wall, in the case of Durham—and would have been behind a person facing an armarium. Thus, to take a book from the shelf and open it while still standing facing the armarium would have put the book in shadow. To see its words more easily, the reader would likely have turned a bit to get the pages of the book into what light was available. The best and most comfortable position might have been to face up or down the arcade, as those sitting in the carrels did, for to face directly outward with a book might well have introduced glare onto its pages.

As long as book collections were relatively small and grew at a slow pace, habit ruled and the little inconveniences associated with consulting a book in one's shadow—perhaps with the top of the generally heavy book resting on the edge of a chest or an armarium's shelf edge—or turning 90 degrees to catch the light just right, and supporting with some difficulty the weight of the book, were not enough to change anything very radically. Thus there was little pressure to relocate book-storage furniture to where it might be better situated in the light—until the number of books in libraries grew to almost unmanageable numbers, which changed not only how they were stored and displayed but also how they were used.

CHAPTER FOUR

Chained to the Desk

When libraries in the Middle Ages had to be moved, they were often transported in the same chests in which they were kept. For monasteries especially, these chests of books continued to multiply. This occurred in part as collections of books from deceased owners like bishops were bequeathed, complete with the furniture that contained them, to monasteries already beginning to overflow, relatively speaking, with books. Keeping all those books secure among populations of monks, and also their guests and visitors to the monastery, created problems of management and convenience, especially if the keepers of the several chest keys had to be assembled every time someone wanted to consult a volume.

Chests were fine for moving and storing books, but they were far from the best way to provide access to them. When books are piled one upon another, many books may have to be moved to get at one near the bottom of the chest. This annoyance could be alleviated somewhat by placing the books next to each other in the chest, with one of their edges facing up, as seems to be the case in the depiction of the abbot Simon. Since there were few identifying markings on individual books anyway, it often made little difference which end was up. If need be, the location of the books in the chest could be indicated by a table of contents attached to the inside of the chest's lid (much as is done in boxes of chocolates today).

An armarium may be thought of as a chest left sitting on its end, in which case the lid becomes a door. Such was the way some deluxe steamer trunks were to be oriented when used in the heyday of travel by steamship. But just turning a trunk on its end would clearly have jumbled its contents. The steamer trunk thus developed into a rather sophisticated wardrobe, with wires, hooks, compartments, and shelves designed to keep everything neatly in its place.

Simply turning a book chest on its end would not have worked very well either, for the books would have come to be heaped together or piled unmanageably high. By fitting a series of shelves within the upended chest, however, books could be segregated in manageable piles. Where very valuable books were involved, each might have its own shelf space. To facilitate getting books in and out, wider chests or armaria would have been developed. Wider chests would mean a wider door, which might require an inordinate amount of floor space to swing open. Thus armaria were fitted with double doors, and the resulting piece of furniture might be considered as two upended chests set side by side.

Armaria fitted with shelves made it possible to treat books with more care, and made it easier to retrieve a needed book. The armarium was thus more appropriate than the chest for keeping larger numbers of books in or near the cloister where the monks worked with them. As collections of books continued to grow in monasteries, and later in churches and universities, separate rooms began to be devoted to the housing of books, which came increasingly to be displayed more openly while at the same time having to be kept secure. Through the conversion of part of a sacristy into a library room, for example, books could be displayed openly on tables or in unlocked or even doorless armaria kept behind a single locked door opening onto the cloister walk. The monks were allowed to consult and read the books within the confines of the library, or perhaps throughout the cloister, and a librarian was in charge of caring for the books and accounting for them. But the system was not foolproof, for volumes on occasion disappeared. To prevent such losses, in the Abbey of Evesham, in Worcestershire, it became the custom that the librarian not only cared for the books in the armaria (which increasingly came to be known as presses), but also policed the cloister and kept track of them:

> It is part of the precentor's duty to entrust to the younger monks the care of the presses, and to keep them in repair; whenever the convent is sitting in cloister, he is to go round the cloister as soon

> as the bell has sounded, and replace the books, in case any brother through carelessness should have forgotten to do so.
>
> He is to take charge of all the books in the monastery, and have them in his keeping, provided his carefulness and knowledge be such that they may be entrusted to him. No one is to take a book out unless it be entered on his roll; nor is any book to be lent to any one without a proper and sufficient voucher, and this too is to be set down on his roll.

Even with such concerns and oversight, those institutions and individuals who owned a number of books—nearly all of which, before the advent of printing, could be considered rare—were not averse to showing them off if their display could be managed in a secure way. Books of special value or significance had long had elaborately decorated covers, as can be seen in the oldest known illustration of a book armarium. The picture forms the frontispiece of the *Codex Amiatinus*, and is believed to date from the middle of the sixth century of the Christian era. It shows Ezra, the Hebrew scribe and priest, writing before an open book cabinet. Inside are five shelves, all but the bottom one containing two books each. The books are bound in crimson and lie side by side. Clearly visible clasps across the fore-edges of the books show them to be arranged with their front covers up and their bottoms out.

Ezra's armarium contains nine volumes, with what appears to be the place for the book he is writing in occupied perhaps by a case for reed pens and an inkhorn. The scribe is sitting right next to the book chest, so its doors can remain open without fear of one of its books disappearing in the hands of an unauthorized borrower. Although the shelves in the cupboard are horizontal, they appear to be slanted up toward the back, an illusion created by the fact that the now-conventional rendering of perspective had not yet been mastered fully. This is confirmed by the appearance of the small table nearby, whose left rear leg seems to have given the artist some trouble.

By the end of the first millennium of the modern era, one of the larger libraries might hold as many as several hundred volumes, and so keeping a book in a fixed and predictable location became increasingly important. Assuming Ezra's book chest was typical in size, capacity, and arrangement of books, a library would require such a

The fourth-century B.C. *scribe Ezra is shown here working before an open armarium in this frontispiece to an* A.D. *sixth-century manuscript.*

piece of furniture for every ten or so volumes, and the floor space required would be proportionately large. If it is estimated that Ezra's armarium required about 5 square feet of floor space, then the ten armaria needed to house a library of one hundred books would cover 50 square feet. Arranging that many book presses in a single room would not be a trivial exercise; since there would have to be ample empty space around them for the doors to be swung open and for readers to move about and read, the total space requirements of the room might be as large as 100 or 150 square feet, or more, depending on how the armaria were placed. This is about the size of the book room that was created at the end of the sacristy opening into the cloister at the Abbey of Fossa Nuova, near Terracina, in central Italy. The room was very near the *armarium commune,* or common press, that was built into the wall beside the door to the church and in which books used in the services were kept.

With a separate lockable room to hold books, the natural evolution toward the more efficient bookcase can be imagined to have proceeded by the doors being taken off armaria of the kind shown with Ezra, which enabled them to be placed closer together in the room and thus accommodate more books. But open chests would have left the books exposed and vulnerable to theft—or to the lesser crime but nonetheless inconsiderate act of a borrower returning a volume, if it was returned at all, to other than its rightful place.

Books generously decorated with precious stones and precious metals would most likely have been kept in more secure armaria and not shelved with the more common books. Some of the heavily decorated book bindings of the Middle Ages were as potentially harmful to other books as was a knight in studded armor to unprotected foot soldiers. Indeed, the juxtaposition of certain books was cautioned against as late as the mid-nineteenth century, with the obviously needed warning that "Books with clasps, bosses, or raised sides, damage those near them on the shelf." It was advised that "books with carved bindings or with clasps should be kept in trays, table-cases, or drawers, not on shelves, for the sake of their neighbours." Keeping bossed and studded books individually face up on tables or individual lecterns also prevented them from harming their neighbors, but as the number of books grew and questions of security increased, alternative ways of shelving them were needed.

Simply crowding more and more armaria into a room like crates in a warehouse would not do, for the tallish structures would begin to obstruct one another's light and conceal the malicious acts of those who might mutilate books by, say, removing the margin of a page for a piece of parchment upon which to write some notes. One way to accomplish the technical objectives—displaying the books while not at the same time obscuring light or secreting the reader from view—was to arrange the books not in compartmentalized armaria but out in the open on long, wide lecterns situated in a special room, much as pews might be in a church. This is in fact what came to be done, and the lecterns had sloped surfaces upon which the books could be displayed side by side. With the lectern at a convenient height and angle for a reader to stand or sit before it, any book could be opened and consulted where it lay. To assure that the books were not removed from their rightful lectern, they were chained to it. This constraint led to other developments, for,

> A chained book cannot be read unless there is some kind of desk or table on which to rest it *within the length of the chain;* that fact conditioned the structure of the bookcase. Again, since a chained book cannot be moved to the window, the window must be near the book; that determined the plan of the building.
>
> Chaining, then, in ancient libraries is not an interesting irrelevance. The fact that some anthropoid ancestor began to employ his front paws for grasping instead of for walking conditioned the upright posture of man and his use of tools—and so his whole future development. Just so, the fact that books were chained conditioned the structure and development of the historic English libraries to the end of the seventeenth century; and it did this even where, as at Cambridge after 1626, chaining began to be disused. Books continued to be chained much later than is commonly realized. Fresh chains were being purchased at Chetham College, Manchester, in 1742, and at the Bodleian in 1751. At The Queen's College, Oxford, the chains were not taken from the books till 1780; at Merton not till 1792. Magdalen was the last college in Oxford to retain them; here they lasted till 1799.

The practice of chaining books thus has a long history. Among the first implications of chaining was to obviate the need for the constant availability of keys to unlock rooms, chests, and armaria. The books were openly available but secured by chains that ended in rings strung on a long rod, as shower curtain rings are on a shower rod. Because the books had to be read "within the length of the chain," they were stored side by side on the same lecterns or desks to which the rods were attached. When the lecterns all faced one way, a series of them might be arranged like pews in a church, often complete with seats at which the readers sat. Other arrangements included double-sided lecterns, placed back to back, with back-to-back or shared benches between them. Some lecterns were built chest high, thus eliminating

This detail of a lectern in the library at Cesna shows the books chained to a rod located below the lectern. Books not in use on the lectern could easily be stored on the shelf below it.

the need for seats and thereby increasing the floor space available for more lecterns and hence more books. (Some modern libraries, such as the State Library of Victoria, in Melbourne, Australia, are fitted with such lecterns over reference shelves, but with the chains made unnecessary by electronic gates through which patrons of the reading room must pass.)

The strong iron chain securing a book to a medieval lectern was just long enough not to impede a user from opening the book and reading it. When the book was not in use, it lay cover up on the lectern, as if on display. Depending upon whether the rod was located above or below the lectern, the chain would likely be attached to the top or bottom of one of the book's covers, which typically were made of a relatively heavy slab of wood, perhaps ¼ to ½ inch thick, depending on the size and weight of the volume and strength of the wood. Chains were also attached to the sides of book covers, near the clasps. Some early rods might have been wooden dowels, but these would have been easily worn out or broken and thus not have provided security. Hence, iron rods soon came to be employed, especially in more heavily used libraries.

At least one student of the practice of chaining books has wondered, "Is it a desecration to assail these venerable libraries by methods of structural analysis and historical research—to view them as a series conceived in evolutionary terms?" Not everyone has viewed them so, and a more poetic than scientific explanation of the origin has also been offered:

> Books of controversy, being of all others haunted by the most disorderly spirits, have always been confined in a separate lodge from the rest; and, for fear of mutual violence against each other, it was thought prudent by our ancestors to bind them to the peace with strong iron chains. Of which invention the original occasion was this: When the works of Scotus first came out, they were carried to a certain library, and had lodgings appointed to them; but this author was no sooner settled, than he went to visit his master Aristotle; and there both concerted together to seize Plato by main force, and turn him out from his ancient station among the *divines,* where he had peaceably dwelt near eight hundred years. The attempt succeeded, and the two usurpers have reigned ever since in his stead: but, to maintain quiet for the future, it was decreed that all *polemics* of the larger size should be held fast with a chain.

Regardless of which account is believed, the chaining of books to medieval library furniture created no small inconvenience. If a monk wished to remove a book to copy it in his carrel, or if another monastery was granted permission to borrow a book from a chained library, the fixture holding the iron rod in place had to be unsecured and all the chain rings removed until the one associated with the desired book was reached. The book could then have been taken from

A chain ring attached to a cover board is seen in this detail, which also shows how the book is held closed with a clasp, which sometimes held marks identifying the book. Chains were attached at various locations on books' covers.

the lectern, and all the other chain rings would have had to be replaced on the rod in their original order, lest there be tangled chains and confusion about a book's location. The process of unsecuring the chain rods was highly regulated by locks and keys (just as multiple keys were required to open book chests, not uncommonly at least two keys held by two different persons were required to release a hasp or hasps on the end of the lectern that secured the rods in place).

Two basic kinds of lecterns were those at which scholars stood and those at which they sat. The standing lectern, like those at the State Library of Victoria, was more common at Cambridge University and remained popular there well into the seventeenth century. In some library rooms at that institution "there is no evidence that any other kind of desk or table was originally provided." Some colleges had lower lecterns, before which the scholar sat—as in Peterhouse, which in the year 1418 had a total of 302 books, 143 of which were chained and 125 were "assigned for division among the Fellows." The remaining books were described as "those of which some are intended to be sold, while certain others are laid up in chests."

It is not clear that the standing lectern is older than the lectern

Two distinct keys were clearly required to release this hasp on the end of the book presses in the library at Trinity Hall, Cambridge, which was completed around 1600. With the hasp freed, the chain rod could be withdrawn just enough to remove or add chain rings.

before which a bench was located. In fact, the overwhelming predominance of evidence for the use of the latter strongly suggests that it was the design of choice and the one that was developed first, perhaps from the church pew. In a pew, a monk could sit and rest a hymnal or psalter on the back of the bench in front, and at a convenient angle. Just as it made long chapel services more physically tolerable to be able to sit or kneel during part of the service, so it would certainly be more conducive to working long and hard in a library equipped with seats before the lecterns.

The standing lectern might have evolved from the sitting one as a space-conserving measure, with the space that benches occupied being used for more lecterns and hence more books. On the other hand, it might also be the case that the standing lectern evolved first, being suggested to monks who experienced discomfort holding a

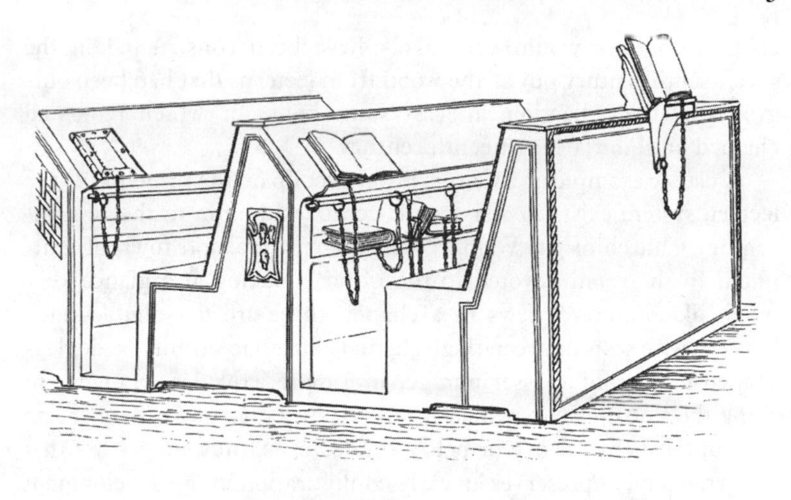

Though most lecterns in the library at Cesna are designed for seated readers, some are for standing readers.

heavy volume while standing during part of the religious service. Given the variety of lectern systems that developed, the sitting and standing lecterns might have evolved simultaneously and independently at different monasteries. (It has even been suggested that the book lectern evolved from the prie-dieu, at which monks knelt to pray.)

In any case, many a library room set up with book lecterns could easily be mistaken by a modern visitor as a chapel fitted with pews. Indeed, when we sit in a church pew we often find books—hymnals and psalters—stored on the back of the pew before us, and the pews and choir stalls in some chapels are even fitted with lectern-like desks on which service books may be placed. In the earlier part of the twentieth century, in an odd twist of artifactual evolution, it was discovered that pews at Hereford Cathedral had actually been made from old lectern seats that had been removed from the cathedral's library when it had been renovated in the previous century. A similar thing happened at King's College, Cambridge, where a College Order of 1851 recorded an agreement with a carpenter named Mr. Rattee "to convert the materials of the Bookcases in the Side Chapels into Seats with Book Boards and Kneeling Stools for the Chapel." The bookcases themselves would very likely have been constructed in the seventeenth century out of the wood from lecterns that had been outgrown by the collection, at least some books of which remained chained until the late eighteenth century.

A classic example of a Gothic library set up on the principle of the lectern system exists in a sixteenth-century addition to the twelfth-century Church of St. Walpurgis in the east Holland town of Zutphen. In the room devoted to the library, looking at a glance very much like a row of pews in a chapel, there are ten double-sided lecterns with seats between them aligned along one side of the dogleg-shaped room, and a lesser number in a more irregular arrangement along the side opposite, which is pierced by a doorway. There are no horizontal shelves above or below the lecterns, thus suggesting that the arrangement preserves an early configuration in the development of the bookcase. The seats are plain benches, with only some modest decoration at their ends to distinguish them from benches that we might see today at a Little League baseball field or in a common locker room.

Though the orderly library at Zutphen does represent a fine example of the lectern system, whether it is the way the earliest lecterns were arranged or, as has been suggested, a latter stage in their evolutionary development can probably not be determined with any certainty. Most medieval libraries set up on the lectern system did, in fact, have more refined back-to-back benches or seats, with one facing each lectern. The Zutphen arrangement, which occupied less floor space, may have been less directly related to the way private studies and monk's carrels were set up, in which a single bench or seat faced a single lectern. Since, unlike the pews in a church, library lecterns did not have to face in a single direction, the arrangement of the furniture could be dictated not by religious considerations but for the convenience and efficiency of installation. Back-to-back lecterns and shared benches allowed more books to be displayed in a given floor space, as at Zutphen.

Because books were chained to the long lecterns and thus had to be read at them, the availability and quality of light was of utmost concern. Readers and keepers of books knew that having one's back to the light was not good, and turning to the light with a chained book in one's hands was not practical, or in most cases even possible, given that the chains were usually only long enough to allow the books to be

Lecterns in the library attached to the twelfth-century Church of St. Walpurgis, in Zutphen, The Netherlands, have books chained to rods above the lecterns.

used at the lectern. Thus lecterns oriented parallel to windows were undesirable, and as a result the desk surfaces tended to be arranged with their long dimension perpendicular to windows so that daylight would illuminate the books from the side. The library at Zutphen is irregularly shaped because the room bends to conform to the apsidal church geometry, but to the extent possible the lecterns are arranged to take advantage of the light.

In fact, wherever the lectern system was employed, whether for standing or sitting readers, its furniture was arranged as much as possible to use the light coming through windows. When preexisting rooms, whether side chapels or old halls, were converted to libraries as growing book collections demanded, one generally had to take the window arrangement as it was found. The buildings were very likely, after all, made of stone, with walls that were structurally significant.

Windows could not easily be moved at will as they might be in a modern curtain-walled building, where the whole wall could be one large window anyway. In the Middle Ages, when lecterns and seats were put in place in a room not designed to be a library, some benefited from being beside a window, though perhaps not located at the ideal height on the wall, while others were necessarily butted up against a long wall between windows. The light available for reading a book chained at a lectern thus ranged from excellent to poor, depending on its location.

Whenever a new room or building was constructed specifically to house the growing number of books owned by an established institution, problems with light were of prime consideration. The library room was often built over an existing structure, such as a cloister walk, which was typically long and narrow. The location of a library on an upper story assured increased security and greater light, but the windows could only be spaced so close, because the wall between the windows was structurally necessary. This fact might indeed have encouraged the development of back-to-back lectern arrangements. Given the structural constraint, the spacing of the windows could be arranged so that back-to-back lecterns—either existing ones or ones that would be made to order—would just fit between two windows with one end of the lectern butted up against the wall. Seats were fitted between lecterns, with the width of the window just right for accommodating the back-to-back seating at the appropriate distance from the lecterns.

Because of the arrangement of its windows, a purpose-built library room can be identified from the outside of a building as the one with the relatively narrow and regularly spaced windows. (This remains today the most distinguishing feature of an older library building, for a wall of many closely spaced windows usually gives away the location of the bookstacks inside.) The effect was especially striking in buildings that had larger, perhaps Gothic windows and doors on the first floor, such as the Collège de Navarre, which is now part of the École Polytechnique, in Paris. The building was taken down in 1867, but a photograph of it survives. Another example is the library of Merton College at Oxford University, whose regularly spaced second-story library windows can be seen coming together in a corner from two sides of the building when viewed from Mob Quadrangle. One side holds the old library and the other the new, in "old" and "new" parts of the building, respectively, but both have characteristic fenestration. Other English examples include the libraries at Lincoln, Salisbury, St. Paul's, and Wells cathedrals.

Windows and natural light were also important because of the fear of fire, and many old libraries were open only as long as the sun was up, because any use of candles or oil lamps put the book collections too much in jeopardy. When a new library was constructed as a freestanding building, if at all possible it was located far enough from existing buildings so that should a fire begin in one of them the flames could not easily leap the distance to the library. According to a seventeenth-century description of the "old library" of the Sorbonne, "that it might be the more safe from the danger of being burnt, should any house in the neighbourhood catch fire, there was a sufficient interval between it and every dwelling house."

As new books were acquired and a library's volumes multiplied, the lectern system would come under increasing strain, for there naturally would come a time when an entire room devoted to the library would be filled to the floor's capacity with lecterns, and the lecterns themselves would be filled to capacity with books. (Were the 17 million books in the Library of Congress at the end of the twentieth century stored on lecterns, they might require as much as 2,000 acres, or 3 square miles, of floor space. The Library of Congress would sprawl across the Mall, spill over onto all the surrounding land now occupied by the Smithsonian Institution and the Washington Monument, and reach all the way to the White House.) With the books so crowded together, there would no longer be enough unoccupied space on the lectern to open wide a volume that one wished to consult. We can imagine that to make room a chained volume not being used might be hung over the edge, dangling from its chain not unlike the way a telephone book is often seen doing in a public phone booth. This would certainly not have been an aesthetically pleasing solution, and it was potentially damaging to the book's binding. As such it could not have been considered good library practice.

Lecterns were used not only to display elaborately bound closed books with pride and care but also to hold them in a convenient position for reading and consulting. In addition, the medieval lectern doubled as a desk on which to place the sheet or book in which the

The lecterns in the library of the fifteenth-century pope Sixtus IV, as shown in this Roman fresco, were filled to capacity. The artist is believed to have left out the details of the chains.

scribe or scholar might be writing. Indeed, the term "desk" is often employed to describe the sloped lectern surface. On a lectern loaded to capacity with chained books, however, it might be very difficult to move volumes aside to make desk room, especially as the chains were often not much longer than they had to be to secure a volume in place. There was, in short, not much room to maneuver the existing books to make a work space on the sloped desk, and generally not a convenient horizontal surface on which to place securely one's inkpot.

Such limitations and annoyances of the lectern system (as it has come to be called by historians of libraries) led it to evolve additional features, which in turn caused book furniture to become more capacious. The first step in the evolution toward the bookcase as we know it and use it was for a horizontal shelf to be added atop or below the inclined lectern. On this shelf could be placed some of the books that were crowded together on the sloped surface, thus freeing up working space. The horizontal shelf, especially when it was above the lectern, also provided a convenient place to set down an inkpot, thus freeing

The Old Library at Lincoln Cathedral was housed in an early-fifteenth-century timber structure built above a stone cloister. The lecterns were fitted with shelves both above and below the reading desk.

up the hand that was not holding a reed or quill to turn pages or hold them down. Lecturers today appreciate such a place on a lectern to set a glass of water.

The addition of a shelf to the lectern also provided another essential feature in the development of the furniture of libraries, whose collections of books were growing at increasing rates (and at an even

The lecterns in the Medicean Library in Florence, which opened in 1571, in time became very crowded with books. Note the protective sheets that could be draped over the books, and the tables of contents of the lecterns posted on their ends.

more accelerated pace after the introduction of printing with moveable type). New acquisitions could be chained in their appropriate place between existing volumes, but they need not have taken up additional space on the lectern proper because they could be placed in piles, with the less elaborately bound books stacked up one upon another atop the horizontal surface. This worked fine for the storage of the more common books, but it presented problems when they were used. As can easily be imagined, moving one book from the sloped surface to make room for another that was under a stack of books could lead to an intertwining and entanglement of chains—an impossible situation for someone not clever at undoing knots or untangling harnesses. Even if there was no chain reaction of frustration, the chains in time might become so twisted that they were noticeably shortened—like a twisted telephone cord—so that the book to which they were attached could not be brought far enough down from the shelf to be placed properly on the lectern. To address the problems of twisting and shortening, book chains came to be fitted with swivel links that obviated most of these annoyances.

As collections once modest in size grew (as they seem invariably to do), more than one cabinet, lectern, or library room was required to house them, and more space had to be found to hold the furniture. If a room that was designated a library came to be filled with lecterns, the only alternatives were to expand the lecterns into another room or to modify them by building upward. It was the latter solution that was commonly adopted around the sixteenth century.

CHAPTER FIVE

The Press of Books

In institutional libraries, the medieval lectern system of storing and displaying chained books evolved for the same reason that all technology does—those using the system became frustrated with it in a changing environment and felt there must be a better way. In the case of the lectern, whose use had spread from monasteries to universities, librarians and readers alike found increasing fault with it. Librarians, whose charge it was to store and protect books and make them accessible to readers, saw their libraries becoming increasingly crowded. Attempts to expand the library into new quarters must have met with the same problems that requests for more space meet with today. If and when the authorities who controlled space were convinced that the library did indeed need a new room, if not an entire new building, the problem of finding the space or the resources to create and furnish it must often have delayed steps to achieve the goal.

In the meantime, libraries had to make do with the lectern system in place. At first, they most likely squeezed new books—face up—in among the old on the inclined surface on which they were displayed, but this must have created inconveniences for readers, especially if two readers wanted to use books that were adjacent to each other. The unyielding chains would not have allowed the books to be moved very far from their assigned position on their assigned lectern. In time, the books would have become so crowded together that there was hardly enough room to open a volume without placing its open cover atop another.

When Humphrey, Duke of Gloucester was petitioned by Oxford University in 1444 to help with the building of a new library, the problems of overcrowding in the old library room were dwelt upon. According to the petitioners, "should any student be poring over a single volume, as often happens, he keeps three or four others away on account of the books being chained so closely together." The situation seems not to have been much different from what would later develop with regard to patrons consulting books in crowded stacks or with regard to seating space in large reference rooms such as that in the old British Museum Reading Room or the main reading room of the New York Public Library. (It was also to be repeated in the late twentieth century with regard to access to computer terminals in libraries and schools. These terminals, often secured with a cable or other chain-like restraint, required space for their ancillary hardware, plus some additional desk room for a mousepad as well as for manuals or books and paper. As computer use grew, it was not generally possible to squeeze more computers in the same space, and so new rooms had to be captured. In libraries this was often done by removing books to remote storage areas or by replacing bulky reference volumes with compact disk equivalents. As computer use in libraries continued to grow, especially as stacks and towers of CD-ROMs containing encyclopedias and other massive databases vied for space, and e-mail and the Internet became communications media of first resort, institutions wrestled with how to establish policies regulating terminal time.)

Even if fifteenth- and sixteenth-century librarians could secure more space, it must have been clear at least to some of them that the new space too would soon be filled to capacity. The installation of a single horizontal shelf above or below the chain rod and lectern, if not in both places, was a quick but temporary solution, and it was likely to create new problems that were as annoying as the ones it solved. While books of lesser value or bound in more benign covers could be stacked flat on top of one another on a shelf above the lectern, the growing pile must have led to a lot of chain-tangling and confusion as to which side of a pair of back-to-back lecterns a given volume belonged. This would have wreaked havoc with the order in which librarians had arranged the books, and it would have caused situations in which a book that could not immediately be found had been removed from its place to the lectern on the other side.

There also must have been a lot of ink spilled over onto the desk on the other side of a lectern pair, for we can imagine what would have happened to an inkpot sitting over the workplace of a scholar when a reader on the opposite side of the lectern pushed some books back on the horizontal shelf. Many librarians and library users must have thought to themselves, There must be a better way.

The better way came in the form of what has come to be termed the "stall system." According to John Willis Clark, who first looked in detail at the history of the care of books, the lectern on which the titles were displayed and read could not be dispensed with as long as books were chained, which they continued to be in some libraries well into the seventeenth century and, as noted, in some cases into the late eighteenth. The problem then was to maintain the desk with its chains but to introduce a new element into the piece of furniture to which they were attached. This was achieved, according to Clark, by an imaginative carpenter who is said to have reasoned that a useful arrangement could be had "if the two halves of the desk were separated, not by a few inches, but by a considerable interval, or broad shelf, with one or more shelves fixed above it." The creative carpenter's plan would lead to what we today call a double-wide case of bookshelves between and above the separated lectern desks. The arrangement is something like a pair of modern secretaries back to back with their desks open.

Canon Streeter, who took issue with Clark's explanation of the origin of bookshelves, believed that the modern furniture of libraries evolved by a combination of the lectern with the armarium. According to Streeter, the bookshelves of the sixteenth century owed their characteristics to the armaria of the Middle Ages, many of which were divided horizontally as well as vertically and thus formed pigeonholes to keep the books separated from one another, lest "they be packed so close as to injure each other or delay those who want them." Regardless of how the bookcase came to be added between and above back-to-back lecterns, it was a pivotal step in the evolution of the bookshelf and how books are placed upon it. Streeter dates the innovation to the late sixteenth century.

In time, horizontal shelves with vertical ends came to hold books in a vertical position, something that was not generally done prior to the stall system but which became commonplace with it. It is likely that at first books continued to be laid horizontally on the shelves, as they were in armaria. Indeed, the higher of two shelves may at first have been intended to have still-chained books piled upon it as they were removed from atop a book on the lower shelf. In this scenario, getting to a desired book would have been like getting to a large serving platter stored beneath a stack of dinner plates in a china cabinet, or maybe even a bit like playing the Vietnamese game "Towers of Hanoi," in which different-sized disks on one peg are to be relocated in the same order on a second peg (but without the maddening constraint that no larger disk can ever rest on a smaller one). The fact that the medieval shelves were double-width between the desks obviated the problem of pushing books or inkpots off the other side in the process of moving a pile from one shelf to another, but piling book upon book, including larger ones on smaller, no doubt led to some precarious situations.

In time, of course, as more and more books were added to the library collection, the lower shelf would come to be filled, and books would begin to be stored on the upper shelf as well—all still in a horizontal position. As this continued, there eventually would not be enough empty space above the various piles of books to hold the books that one wished to remove from atop a book one wished to consult. To attempt to pull out one book from the bottom of a pile of heavy volumes stacked horizontally atop it was then, as it is now, to tempt gravity even more so than in snapping a tablecloth out from under a table setting.

It may never be known how and when it occurred to a librarian—or perhaps, possibly, to a reader actually wrestling with heavy tomes—to arrange the books vertically on the shelf, which not only gained space for more books but also positioned them so that any one of them might be removed with minimal effort and resistance from between its neighbors, which need not be moved in the process.

The problem with placing a number of books in a vertical position is, as everyone knows, that if they do not fill the shelf from end to end they can flop over unless they are shored up by the last book in the line leaning at an angle, or by a pile of horizontal books, or by the presence of a bookend, which seems not to have been a common object in the Middle Ages or Renaissance. There are depictions of small numbers of books arranged vertically on the shelves of contemporary private studies, but they are clearly exceptions. Indeed, the vertical shelving of books did not come about until they could no longer be accommodated easily in the horizontal position—not, that is, until overcrowding severely taxed the stall system.

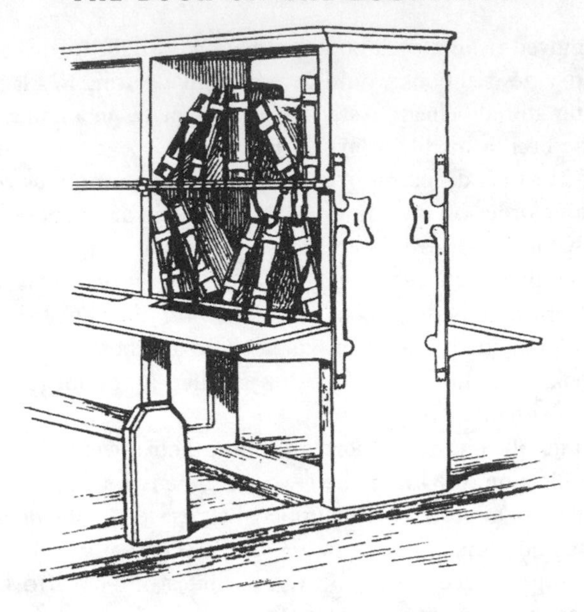

Shelves built above and between back-to-back lecterns provided space to store books not being used on the desk. By shelving the books vertically rather than horizontally, a single book could more easily be taken down from the shelf, with minimal disturbance of the other books.

Scholars or librarians did, in time, get the idea of arranging the books on the shelves above a lectern in a vertical position, and the problem of the books falling over was solved because by this time they were so numerous that they filled the shelf completely from vertical end to vertical end. Even when this point was reached, however, the evolution of the use of the bookshelf as we know it today was still not complete, for the books were at first shelved with their fore-edges out and their spines facing into the shelf. The reason for this—in addition to the fact that there was no need to see a spine that bore no identifying author or title—was that the books were still chained, and the chain could be attached to any of the three edges of the front or back cover of the book but could not easily or effectively be attached to the spine. For example, when books were stored horizontally on a shelf above the desk, the chain may have been attached to the top of the

This detail shows how a vertically shelved book was chained to a press wide enough to accommodate two rows of books. Note the spines of the books chained to the other side of the press, indicating that their fore-edges were facing outward.

back cover. In this way the chain did not deface the front cover, which was the more decorated face of the book, and also the chain was least intrusive when the book was placed on the lectern to be read. Some books had chains affixed to the bottom of the back cover, which would have been more convenient for books shelved beneath the lectern, as they were in some libraries. It appears to have also been a matter of local custom and craft idiosyncrasy exactly where a chain was attached to a book.

When books came to be shelved vertically, connecting the chain to the top of one of the cover boards would have led to it being draped over the fore-edge or side of the book, thus doing damage by getting in between the books or their pages. Attaching the chain to the bottom of a cover board would have caused it to scrape the shelf as the book was moved in and out of its place; it would also have caused the book to sit tilted on the shelf, thus stressing its binding in a harmful way. Furthermore, unless care was taken, either the bottom of the book would sit on the chain, thus resulting in a volume askew and potential damage to its pages and binding, or the chain would become wedged between two books, wearing on both of their bindings. Thus, the best location for the chain to be affixed to the vertically shelved book was a

The hasps on the presses at Hereford Cathedral are long enough to secure the ends of three rods with a single lock. Here a rod is shown partially withdrawn, which allows access to the chain rings.

cover's fore-edge. To allow the chain to hang down in front of the shelf and interfere least with other books, it was natural to shelve the books with their fore-edge facing out. This arrangement can be seen in numerous illustrations in Clark's essay and Streeter's survey, where the preferred attachment point of the chain appears to be the upper part of the front cover.

Thus, the chaining of books, no matter where the attachment was made, conditioned library users to shelve books as much as possible with the chain attachment outward on the cumbersome bookcases. This meant that any edge of the book but its spine faced outward from the shelf. Having the spine face the back of the shelf also came to be adopted in private libraries and studies as the customary way to display even unchained books when, because of their increasing numbers, they too came to be arranged side by side vertically.

The vertical dividers associated with the bookshelves introduced in the late sixteenth century may very well have been suggested by the armaria of early medieval or older times, but it is also likely that they were present for purely structural reasons. Had they not been, the arrangement of books vertically might have taken even longer than it did to evolve, for it is the intermediate vertical supports that not only constrain how many books can be shelved horizontally but also provide the rigid "bookends" that hold a manageable number of books upright.

The length of a typical medieval lectern was in excess of 7 feet, and a shelf of that length supported only at its ends would have sagged significantly, if not pulled itself off of its end supports, especially when loaded with heavy books. Noticeable bowing would have been unpleasing to the eye and possibly harmful to the books if the top shelf deflected down so much that it rested on the books below, causing the further problem of books being damaged by being wedged in so tightly that they could not easily be removed. Thus, the shorter the distance between shelf supports, the better.

To the modern engineer, equipped with theories and formulas unavailable in the Middle Ages, the engineering problem of designing a solid-looking bookshelf is essentially no different from that of designing a bridge. A shelf loaded with books or a bridge loaded with bumper-to-bumper traffic is what is known to engineers as a uniformly loaded beam, and its strength is calculated by a well-established formula which says that doubling the span quadruples the stress the beam must resist, while doubling the depth diminishes the same stress to one-quarter the amount. In other words, as far as strength is concerned, we can achieve the same effect by shortening the length or increasing the depth of a shelf in the same proportion.

Sag, known to engineers as deflection, is another matter, for changing length and changing depth do not have proportionally opposite effects. If we double the length of a bookshelf, it will sag sixteen times as much when fully loaded with the same kinds of books.

This book press in Oxford's Bodleian Library clearly shows the shelves sagging under the weight of books.

Doubling the thickness of the shelf, however, will diminish the sag by a factor of eight. Thus, one way to keep extra-long shelves from sagging excessively is to make them disproportionately deeper, something that goes against most people's notion of what a well-proportioned bookshelf should look like.

According to those who have thought about the question, "Probably the most often seen mistake in the installation of bookshelves is the failure to factor in the sag quotient." The librarian and metrician Melvil Dewey believed that the "golden mean" of bookshelf length is 100 centimeters, or about 40 inches, because of "the experience that shelves over 100 cm long sag in the center from the heavy weight when full of books. This sagging not only ruins the appearance of the library, but sometimes causes the shelf to slip off its supports, and always tends to tip the books in toward the shelf center."

An elementary calculation, which can be made by using a beam formula learned by sophomore engineering students, shows that a simply supported long shelf board—say one 7 feet long, 1 inch deep, and 12 inches wide, the size that may perhaps have been used in the Middle Ages—will deflect more than 2½ inches at its center under a modest load of books weighing 4 pounds per inch of book thickness. A modern 36-inch bookshelf, made of ¾-by-8-inch pine, supported at its ends by pegs and holding a full load of similar books, will have deflection of about ⅙ inch, smaller by a factor of 15 but a still noticeable amount. Shortening the shelf by only 6 inches makes a big difference in the deflection, because the sag is proportional to the shelf length to the fourth power. Indeed, the sag of a 36-inch shelf is more than twice what it would be for a 30-inch shelf cut from the same board and filled with the same density of books. To prevent the potential aesthetic and functional failures of excessively sagging shelves, experienced carpenters in the days before such calculations could have been made—days when things tended to be overbuilt, by modern standards—would have supported the shelves every couple of feet or so. The boards used as vertical supports fortuitously would also come to change the way books were arranged on the shelf. Engineers know they cannot escape the laws of nature—whether in seeking thermodynamic efficiency or the perfect bookshelf—but sometimes they can come up with clever solutions that pit nature against herself.

A college classmate whom I once visited in West Lafayette, Indiana, also worried about the strength and appearance of his bookshelves. My friend had just moved into a new studio apartment, which he wanted to subdivide with some freestanding bookshelves. Being a graduate student, he was planning to build them out of bricks and boards. Since he was an engineer, he viewed the bookshelf as a structure, a wooden beam supported by piers of bricks. Just as a highway bridge's roadway should be as level as possible, so my friend wanted his bookshelves to be as straight and level as he could make them. He knew that all beams bend noticeably when a heavy enough load is piled upon them, as he expected would be the case with the engineering textbooks and handbooks he wished to shelve on the unusually long boards that he wanted to use. My friend knew that if he put the bricks under the ends of a board, the shelf would sag significantly in the middle. He could, of course, have added a third pile of bricks to support the middle, but either he did not want to spend the extra money and haul the extra weight home, or he did not desire to give over to bricks space that could be used for books. But mostly, I expect, he hoped for what engineers describe as a more elegant solution.

It does not take an engineer to know that the greater the distance between the brick supports the greater will be the sag of the board. Hence, if the bricks are moved closer together the sag in the middle will be reduced, but when books are placed on the center portion of the board, the overhanging parts will be deflected upward like the wings of a glider in flight, thus leaving a noticeably bowed shelf. If, on the other hand, the books are grouped on the overhanging parts of the shelf, they would bend down like the drooping wings of a glider on the ground. My friend knew he would be filling the finished shelf from end to end with books, however, and so as the volumes in the middle depressed the center of the shelf, they would lift the overhanging portions. Those books on the overhanging parts of the shelf would push it down there, but that action in turn would lift the central part of the board between the bricks and reduce the overall sag there. In fact, using some of the same handbooks that he wanted to support on the shelf, my friend derived formulas and calculated the exact distance along the board that the bricks should be placed so as to minimize the deflection of both the ends and middle of the shelf and make it so that any slight remaining sag would hardly be noticed by the casual observer. An engineer would say that the design is optimized, at least with respect to sag and space for books.

Another way to reduce sag is to attach a relatively deep strip of wood to the front of a thinner board, thereby increasing its depth and stiffening it. For example, the bookshelves in my study are of ¾-inch plywood faced with a 1-inch-deep strip of solid wood, about ¾ of an inch wide, installed flush with the top of the shelf. This not only finishes the plywood as does a veneer but also stiffens it as if it were a deeper beam. Such a solution has the added advantage of giving shelves the appearance of being deeper than they really are, thus enabling thinner or less expensive wood to be used while at the same time achieving the proper proportions of shelves to their span and to the uprights of the bookcase. Samuel Pepys's seventeenth-century bookcases, which still survive in Magdalene College, Cambridge, have brass rods fitted beneath some of their shelves, presumably to boost once-sagging shelves back into a horizontal profile.

In the rare-book stacks of the University of Iowa library there is a collection of books shelved in what appear from a distance to be wooden cases from a private library, because they are topped by a handsome cornice. As used bookstores often acquire the bookcases with the books, so must the special collections of libraries inherit them (much the way medieval monastic libraries acquired along with a bishop's books the chests in which they were kept and transported). The Iowa cases, however, proved to be made of steel finished to mimic wood. A closer inspection of the guts of the cases, where the sheet-steel shelves are supported in slotted pressed-steel standards, revealed hinges, showing that the cases once had doors hung upon them. Some of the shelves are double-wide, suggesting that the original owner wished to store books two rows deep—as Pepys did in the seventeenth century and as this library was now doing in the late twentieth century—or to store large volumes on their sides. Beneath the sheet-steel shelves, which are folded down in the usual way along their front and back edges for stiffness and strength—and not incidentally to give them the proportions of wood shelves—there is also an additional piece of folded sheet-steel welded along the entire length of the shelf, to form a box beam and give added strength and stiffness to the shelf. The owner of these shelves must have appreciated the straight lines they maintained even under the extra-heavy loads to which he must have subjected them. His satisfaction was probably not unlike that experienced by medieval librarians.

The vertical partition boards used in the late Middle Ages to keep shelves from sagging also, perhaps fortuitously, provided a means of conserving space and providing easy access to books by allowing them to be shelved vertically. When institutional shelves were appropriately filled, and the end boards acted as bookends, books stood front cover to back cover and mutually supported each other—not too tightly or too loosely in the ideal situation—in straight-up fashion with their chains hanging down from each of them. With this arrangement, it was easy to remove any one of the books without disturbing the other books or chains. The ensembles arranged in this way would have been straighter and neater than they had ever been, for a horizontal piling does not lend itself easily to keeping a neat arrangement. Indeed, a line of uniformly sized books arranged vertically like soldiers standing at attention would have resembled the order of nothing so much as a stack of books in a bookbinder's press, where they had to be arranged carefully so that the glued binding would not dry askew or the pile buckle under the screw pressure and thus ruin a lot of hard work.

The visual appearance of books standing neatly arranged in a bookcase must have reminded more than one observer of the books in a bookbinder's press. Whether this similarity did in fact encourage the use of the name "press" for bookcase is one of those elusive questions of etymology, but armaria did come to be known widely in English as book presses, and in time the terms "bookcase" and "press" came to be interchangeable. However, no matter what shelving units were called, books continued to multiply, and the storage space for them continued to be taxed. Book presses began to be made three shelves high, and the space below the desk, which was what the lectern had become, was eyed for the further storage of books. While there were some notable exceptions—as in the fifteenth-century library in the northern Italian town of Cesna and in the seventeenth-century library of Trinity Hall in Cambridge, where a horizontal shelf was added below the desk—generally speaking the space below was left open and unused, except by the knees and feet of the scholar sitting in the stall. In time, however, and pressed by growing inventories of books and manuscripts, librarians began to store chests of books in this unused space and, later, arranged books on shelves installed there. It appears that the need for shelf space won out over the desire to keep books from being kicked about and abused.

By turning spines outward on the lower shelves, a librarian could ensure that the fore-edges of the book pages were better protected from kicks and scuffs from the shoes and boots of students and scholars. Since the spine was the most vulnerable part of the binding, it already suffered considerable abuse. Yet, as the printing press developed and printed books swelled collections, where to put them was a central concern. Few new books, especially those of smaller dimensions, were chained, and this allowed them to be shelved equally easily with spine in or out, though they often continued to be shelved inward by force of habit.

In the 1620s at Cambridge University, when St. John's College (which the diarist John Evelyn thought "the fairest of that University") was built, a new library was fitted with a new kind of bookcase. The room built for the St. John's library was of remarkable size, 110 feet long. At 30 feet wide, however, it was of a familiar scale crosswise, with "each side wall pierced with ten lofty pointed windows of two lights with tracery in the head." This arrangement had worked well for libraries in the past, and because there were few complaints about it, it was adopted here. The width of the room would allow for two

Chained books eventually filled presses to capacity. These late-sixteenth-century presses at Hereford Cathedral are believed to have served as models for the furniture installed in Duke Humphrey's Library in the Bodleian. Note the half-raised desk, hinged so that the lower rods can be accessed.

rows of 8-foot-tall book presses to be installed perpendicular to the side walls, and a wide center aisle in which tables or lecterns could be placed. The presses were located between the windows, which were 3 feet 8 inches apart, enough to allow for seats to serve desks attached to the presses. But since chaining books was no longer necessary or practical with mass-produced printed editions, there was no need to

The chained library at the University of Leyden is depicted in this 1610 print. The books are arranged according to subject and shelved vertically when not in use. Note the closed armarium in the right foreground and other book cupboards along the rear wall.

install desks right at the bookcase. Without desks, there was no need for seats, and so the space in front of the window became free for a lower book press, which provided shelf space for more books.

It is not known whether the windows were deliberately constructed with sills 4 feet off the floor—to allow something of about that height to be placed before them without obstructing the light—or whether the idea of putting "low bookcases" rather than benches in front of the windows came first, requiring the sills to be 4 feet high. However, whichever was the chicken and which the egg, the arrangement worked well for St. John's and increased the book capacity of the library more than 50 percent. The "low bookcases" were actually "originally 5 feet 6 inches high, with a sloping desk on the top on which books could be laid for study," and were modeled on the standing lecterns that were a Cambridge tradition. Their height was low

Chained and unchained books are shown here juxtaposed in a press at Merton College, Oxford. Note that the former are shelved fore-edge out, while the latter are shelved spine out.

enough not to obstruct the light from the high windows, while at the same time being convenient at chest height for holding the books—whether taken from the low cases at which the reader stood or the taller cases behind—while reading them in a standing position. (In time, the standing lecterns were raised above reading height by the addition of another shelf, with the sloped desk surviving atop the cases as a vestigial artifact.)

For those who preferred to sit while reading, "stools were also provided for the convenience of readers." These stools were not fixed

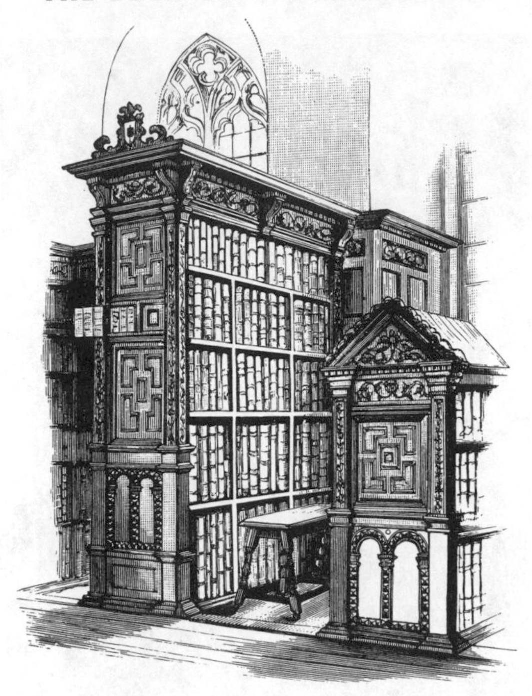

When books were no longer chained, neither desk nor fixed benches were necessary, as illustrated in this arrangement in the library of St. John's College, Cambridge, which was completed in the early seventeenth century. The standing lectern at the right was located directly before the window. Stools were provided as much to enable patrons to reach the higher shelves as to sit before the lower ones.

in place, however, and so could be moved about at will. And they must certainly have had another purpose as well, for their legs were splayed, suggesting that the stools were designed to be stood upon without tipping. This certainly would have been convenient for shorter readers, or for those who wished to get a better view of the books on the upper shelf of the 8-foot-tall bookcases.

It is believed that the original construction of the St. John's bookcases featured a plinth that ran all around the base, resulting in the lowest shelf being raised off the floor. This feature appeared also in

Chains were taken off the books in the library of Peterhouse, Cambridge, in the late sixteenth century. The presses shown here date from the mid-seventeenth century. They are distinguished by the projecting wings at their base, which once formed the ends of benches that extended the length of the press and along the wall between adjacent presses. These seats were also known as podiums, having served as places to stand on to reach books on the upper shelves

the library of Cambridge's Peterhouse, where it seems to have been employed because a bench was fixed to the bookcase on which readers sat to read with their backs to the books. This was possible because there were no chains to restrain the books or in which to get entangled, and the seat surface also served as a place on which to stand when looking at books on the upper shelves. The seat was in time removed, perhaps to make more shelf room available for books, but even many modern bookcases retain at least the semblance of a plinth and a vestigial seat at their base. The fact that such an arrangement and alteration also occurred at St. John's College is supported by the fact that the tall bookcases continued to be referred to as "the greater seats," and the plinth that survives at the ends of the presses shows that, carried around the front, it might have risen even to a height that would have allowed for a back to the seat. With this and the seat removed, new shelves for more book storage space could be installed.

The bookcases at St. John's are also believed originally to have had a pilaster in the middle of the case. Its existence is supported, among other evidence, by the central bracket beneath the cornice, and the pilaster was likely removed when the seat and its back were abandoned in favor of more shelf space, for it would not have been architecturally attractive or structurally true without the plinth or seat to rest upon. The shelves at St. John's are further interesting for the narrow, unadorned verticals that occur on four of the five shelves. The fact that the top shelf has no such verticals constructed on it suggests strongly that the purpose of the verticals was at least in the first part structural—to prevent the sagging of the shelves rather than to provide pigeonholes or lateral support for the books. Today's bookcase would most certainly carry the verticals all the way through to the top, but whether this is done mainly for structural, functional, or aesthetic reasons may certainly be debated.

The vertical elements that prevented sagging also partitioned old bookcases, and in this regard they played an important role not only in holding books in an upright position but also in locating them; the catalog that was posted on the end of a press grouped books by partition, so that they could be more easily found when wanted. In fact, according to Streeter,

> What was added to the lectern was not shelves—conceived of as in modern bookcases—but *partitiones,* or cupboard divisions. This view is further borne out by the importance which continued to be attached to the *partitiones.* This is shown by the system of cataloguing. At Hereford as late as 1749 the catalogues are not based on an alphabetical system; they are tables of contents of the *partitiones* in each several bookcase.

Thus, the unpartitioned top shelf in the St. John's library would definitely have to be associated with a latter stage in bookcase development, for according to Streeter, the bookcases introduced by the stall system were not so much seen as long shelves subdivided by the vertical elements but as a series of compartments that happened to line up horizontally. Indeed, the Hereford bookcase partitions are numbered across rows rather than down columns, lending credence to Streeter's contention that it was the partition and not the shelf that was the primary unit of the bookcase.

Thus, the way books are still located in the restored chained library at Hereford Cathedral—claimed to be "the finest remaining British chained library," with fifteen hundred books "chained to their 17th century book presses"—is as follows:

> Each book, as is customary in historic libraries, has a unique shelfmark representing its precise position on the shelf. This is made up of three elements: a letter representing the bay (A–P), a number for the shelf, and a further number for its position on the shelf. The shelves are numbered from the top and then from the inside of each bay, so that the inner shelves in the three rows of each bay are numbered 1, 4, and 7 from the top: these numbers are still painted on the edge of the face-end on some of the cases.

This is dissimilar from the way our libraries, bookstores, and private book collections are arranged today, and the change likely came when rows of bookshelves began to be extended for considerable distances. We do not follow a shelf past a vertical support when we are browsing for a book, but rather we return to the left end and go down a shelf in that group of shelves—now known by American librarians as a "section" but long called a "tier" in England—to continue the ordered series of books, whether it be arranged according to the subject, the alphabet, or a numerical scheme. In effect, the layout of our bookshelves is now in column form, as was the writing on ancient scrolls, rather than in the long lines of shelves that sometimes form the dominant visual but not ordering element of larger libraries, bookstores, and home studies. Even books themselves, which are read completely down one page before proceeding to the top of the next, echo the arrangement of the modern scheme. We would never dream of reading across the gutter of a book to finish the top line of a right-hand page before returning to the left-hand page to continue with the second line. Another analogy is due to Melvil Dewey (or Dui, in the form to which the spelling reformist and library taxonomist unsuccessfully tried to shorten his name, as he shortened many words). Dewey, who used the British term "tier" for section and "face" for press, perhaps because the words took up less space, wrote:

> The shelf corresponds to the line of a newspaper, the tier to the colum, the face to the page. The invariable library rule should be to arrange books so to follow their order like reading a newspaper,—left to right, top to bottom,—and no more run across an upright than read across the rule between colums in the paper. To number and arrange shelves from bottom up is as Chinese a method as to number books from right to left, or to arrange a card catalog from the back of the drawer toward the front.
>
> This awkwardness arose from shelving high rooms near the floor and gradually adding other shelves above, all being numbered like the stories of a tall house. A theory can be found for almost any practice, but this is so plain an inversion that it should be set aside.

It was principally in the larger institutional libraries, which began to multiply in the sixteenth century, that orderly arrangements of books were necessary not only to economize on storage space but also to help both patrons and the librarians—the latter being often more keepers than daily users of the books under their care—locate individual volumes.

In any case, when books did first come to be shelved in uniform rows in the Middle Ages, the spine was placed inward for the variety of reasons already discussed. In addition, the spine of a book was the "back," the mechanical side of the artifact, not something to display to the world. Indeed, besides being the least appropriate part of the binding to which to attach a chain, the spine of a book might quite likely have been considered the least presentable aspect of it, and as such was to be faced away from sight. The spine was the hinge to the door that is the book's cover, and while hinges are sometimes made to be presentable, they are not intended to be the focus of attention. Door hinges, squeaky ones at least, are annoying but necessary adjuncts to the more significant part—the door. Hinges are to be heard but not seen, perhaps, and in the best of all worlds they are neither to be seen nor heard. The spine of a book was no more meant to be in plain view than was the underside of a table or desk or than is the back of a computer today. (How often do we see in advertisements for new computers the tangle of cords and cables that are so difficult to hide at home?) The book spine provides essential structure, which users of books on tables and desks are unlikely to notice or give a second thought. Of course, spines, like door hinges, do get seen now and then, because books, like doors, must be used, and for this reason the spines of elaborately bound books, like the hinges of ceremonial doors, did get some decoration, but seldom as much as the book cover or door proper. It was only when a book's spine showed unsightly wear and tear that it called attention to itself, and this was another reason for books to be shelved with their spines inward.

In large libraries, books were kept in order, and located with their spines in and with few if any distinguishing features on their fore-edges, by a schedule of a bookcase's contents posted on the end of the case—a table of contents, as in Hereford's chained library. In the stall system of shelving there was typically a wide central aisle down the entire length of the room, with stalls formed by book presses faced by reading desks off to either side. The books in each case were listed in order and posted in a frame installed for the purpose on the end of the press facing the center aisle. Thus, according to a contemporary sixteenth-century description, the large private library of the bishop of Rochester was described as "the notablest Library of Books in all England, two long galleries, the Books were sorted in stalls and a Register of the names of every Book at the end of every stall." If books were exchanged or rearranged, the listing lettered on parchment or paper could be easily updated or revised.

The sometimes elaborately framed spaces for the book list, which can resemble church bulletin boards or hymn or psalm boards, are still attached to the bookcases in many old English libraries. In some cases, the frame in which the list of books was posted was even fitted with a small wooden door or doors that could be closed when the list was not being consulted. This not only added a touch of elegance to the library—as it still does in the Wren Library at Trinity College, Cambridge, where the view down the center aisle of the room is one of wood paneling interrupted only by alcoves of bookshelves—but also the library with covered tables of contents would not have appeared to contain a series of menu boards. Another reason that doors were fitted to the catalog frames might have been to keep the sunlight from fad-ing the ink in which the contents were recorded, for even after printing had been well established as the standard medium for book production, something so singular and also so susceptible to revision as the contents of a bookstall would have continued to be executed in a librarian's hand.

The posting, whether covered or uncovered, of the catalog of a book press on its end has survived to the present time in the labels and signs that are posted on the end of a range of shelves in a library or bookstore. Granted, these do not list all the books, but in a bookstore they often do designate a category, such as history or technology, and in a library a range of call numbers. Because today within the category or range the books are placed alphabetically or numerically, we usually can rather easily find the book we want or conclude that it is not in the collection or immediately available in the appropriate location. Our home bookcases, because their contents tend to be so familiar to us, seldom have or need such rigorous ordering schemes. However, as is alluded to at various places in this book—and dealt with explicitly in the appendix—there are exceptions, and some exceptional book collectors.

Since the location of a book could be determined by establishing its shelf number—by counting down the table of contents, if necessary—finding its position on the shelf, or rather in a partition, was a simple task. And given the locating system, it was not at all necessary to identify authors or titles on book bindings. When there were distinguishing words on a book in a press, they were often written on the fore-edge itself or on the ribbons, clasps, or other devices that held the books closed. (Such closure devices were necessary because parchment leaves would wrinkle if not kept pressed together, and the wrinkled parchment would bulk out the book, causing its fore-edge to swell as much as two or three times the thickness of the spine. With the development of printed books, closing ties were often dispensed with; the paper pages tended to stay more or less flat and compact between the covers, especially when shelved snugly front to back in book presses.) Some chained books were identified by tags attached to the chain itself, a system reminiscent of the tickets that labeled the contents of scrolls.

The bookcase in England was approaching the configuration we now know sometime in the sixteenth century, when the Reformation occurred. The monastic libraries were in effect "the public libraries of the Middle Ages," and the larger religious houses were the centers of culture and education at the time. It was in these, for example, that children were schooled and prepared for the universities. In the three short years between 1536 and 1539, however, "the whole system was swept away, as thoroughly as though it had never existed." As the hostility directed against the clergy by the Huguenot movement in France "exhibited itself in a very general destruction of churches, monasteries, and their contents," in England there was "the suppression of the Monastic Orders, and the annihilation, so far as was practicable, of all that belonged to them." The Protestant Reformation of the sixteenth century created a great disruption, to say the least, in the development of libraries and their furniture.

It has been estimated that "upwards of eight hundred monasteries were suppressed, and, as a consequence, eight hundred libraries were done away with, varying in size and importance from Christ Church, Canterbury, with its 2,000 volumes, to small houses with little more than the necessary service-books." After such destruction, in 1540 "the only libraries left in England were those at the two Universities, and in the Cathedrals of the old foundation." Whereas during the French Revolution of 1789 the books from pillaged convents would be sent to the nearest town, during the Reformation in England no organized attempt was made "to save any of the books with which the monasteries were filled." Furthermore, "the buildings were pulled down, and the materials sold; the plate was melted; and the books were either burnt, or put to vilest uses to which waste literature can be subjected." According to a contemporary report, such uses included tearing pages out of manuscripts in order to have something in which to wrap food or with which to clean out candlesticks and polish boots. Some books were sent by the shipload to other nations, whose gain was, of course, England's great loss. Manuscript pages were employed as endpapers or compressed into pasteboard for covers of early printed books.

The invention of printing itself might have been responsible in part for the fact that only a limited number of manuscripts survived "to give us an imperfect notion of what the rest were like." Because of the proliferation of printed books—many of the earliest of which found their texts in manuscripts and thus multiplied in a single press run a hundred- or thousandfold what scribes had labored over for months, if not years—the value of manuscripts, for their content at least, was considered small. In addition, there were few private book collectors at the time, and so even at bargain prices there was no market.

The university libraries fared little better. Oxford's Bodleian Library, which was completed around 1480, had a considerable number of manuscripts by the middle of the sixteenth century, the most important of which were about six hundred donated by Humphrey, Duke of Gloucester even before the construction of the building. After Edward VI sent his royal commissioners to Oxford (and to Cambridge also) in 1549 to reform the libraries, only three manuscripts were permitted to survive. Few of those that perished were in fact theological literature, and many "had nothing superstitious about them except a few rubricated initials," which made them look like religious works.

So many books were destroyed that the bookshelves were, of course, empty and superfluous. Members of the Oxford Senate were appointed "to sell, in the name of the University, the bookdesks in the public library. The books had all disappeared; what need then to retain the shelves and stalls, when no one thought of replacing their contents, and when the University could turn an honest penny by their sale?" In other libraries, where some books did remain or where there was the intention to refill the shelves with printed books, there was no rush to sell the furniture, but it would be a long time before there would be any need for more shelves or a rethinking of how best to make use of space that was overcrowded with books. In fact, "nearly a century passed away before any novelty in the way of library-fittings" was to appear. In the meantime, the old bookstalls, typically comprising a couple of shelves above a desk, with a fixed seat, continued to serve as the standard design, even though it was the chaining of books that had given the arrangement the configuration that it had. It is such habit that shapes much of our artifactual world, and it changes in time not because its form is antiquated but because it becomes inadequate. In the case of bookcases, this time would come only when the bookshelves had filled up again to the point of overflowing.

During the Reformation, the books of the monastic libraries that survived one way or another were redistributed, and subsequently the great private libraries began to grow. In time, when newer libraries were contemplated and built in locations not smothered in the tradition of an Oxford or Cambridge or other medieval institution, they generally adopted what have come to be familiar to us today as full bookcases without the attached desks, which became unnecessary as books were increasingly more and more available and hence less and less frequently chained.

CHAPTER SIX

Studying Studies

When we sit in a comfortable chair with a book, we tend to hold it open in a plane that makes the words at both the top and bottom of the page more or less equidistant from our eyes. Unless we hunch directly over it, the book laid flat on a desk is not so easily read, for as we progress down the page, the text comes increasingly closer to our eyes. This is no great problem at the rate at which most of us read, but there can be the slightest moment of hesitation when we turn to the next page, because our eyes have to refocus on a more distant line of type. The phenomenon was brought home to me with great force some years ago when I was reading *The Modern System of Naval Architecture,* the magnum opus of the Victorian naval architect and engineer John Scott Russell. This multivolume work has pages measuring 20 by 28 inches, with lines of text that extend across the entire page. Although it is printed in an appropriately large-point size of type, I found the book very unwieldy to cope with. It was too large to read in my carrel and too heavy to hold unassisted in a conveniently slanted position. When I laid it flat on a library table, I could not easily read from the bottom of one page to the top of the next: I had to stand before the table and look directly down on the book to read it comfortably. Had the book been chained to a medieval lectern of a sufficient size, it would have been so much easier to deal with.

To make it simpler for us to peruse even conventionally sized volumes, we have book stands and easels to hold books on our desks or beside the word processor. (The laptop computer itself has a screen whose inclination can be adjusted to suit viewing conditions.) Illuminated manuscripts and portraits of medieval scholars in their studies

This fifteenth-century scholar has a second lectern mounted above his lectern-desk. There are no bookshelves in evidence, but there is a book chest within easy reach.

show books propped up at a readable angle on top of and against other books, against a wall, upon a lectern that rests on or is fitted to a flat desk, or on a desk that itself is slanted. In time, desk lecterns evolved to double-sided lecterns, so that a scholar could hold two books open for study or comparison—one on either side of the desk or table. But to consult the book opposite the working surface, the scholar would have to exchange the places of the books or get up from a seated position and walk around the desk to consult the second volume—unless the lectern could be physically lifted up and turned. Such inconveniences led naturally to a desire to have the lectern itself turn easily, and the lectern came to be fitted upon a post, to bring the desired book into view. Subsequent developments included making rotating lecterns that could hold more than two books; some were fitted on screws much as is an organ stool so that the height could be adjusted.

This woodcut shows Isotta Nogarola, a fifteenth-century scholar, using a rotating lectern in her book-strewn study.

Other, smaller lecterns were fitted on angled-bracket swivel devices in order that the book being consulted could be moved in and out of the working area.

Decisions about how and at what angle to arrange a book to read or even a sheet of paper on which to write, and the choices among pieces of furniture to aid in such arranging, are nothing new. By around the year 1500 the portable sloped-top writing desk, complete with compartments and small drawers suitable for holding writing materials, was in wide use. A dialogue for Tudor schoolboys, in which a scholar modeled after Pliny the Elder directs the arrangement of his working area, indicates that the device provided a convenient addition to the flat desk, and an alternative to the rotating lectern:

PLINY: Set the table up on its supports in the bedchamber.
CELSIUS: Do you prefer the table to the desk?

PLINY: At the moment, yes, but place the small desk on the table.
EPICTETUS: The fixed or the revolving one?

The need of a scholar to have open several books at the same time—or rather the frustrations scholars would have experienced and expressed about using existing desk lecterns that held only one or two books—led Renaissance craftsmen, inventors, engineers, and scholars themselves to come up with increasingly ingenious devices for holding and storing books. Contemporary illustrations show a staggering number of these inventions, from which it seems safe to conclude that questions of where and how to work with books in carrels, private studies, and libraries had come to be a serious topic of thought, conversation, and construction.

New College, which was founded at Oxford in 1379, had a curious room designed to accommodate four Fellows of the college. It was fitted with four windows located much closer to each of the corners of the room than classical architectural principles would have dictated. William of Wykeham, who was responsible for the layout, had a functional purpose in mind, however, and this was made evident when beside each window was installed a desk and a stool, forming a carrel-like space in which the Fellows could study the books assigned to them. This arrangement in the shared room was designed to give more privacy to each Fellow than he would have had if the study spaces had been located beside windows closer to the center of the outside walls.

The Fellows of New College were assigned books for private study because it was inconvenient for the reader and librarian alike to keep volumes that were frequently consulted, whether in a monastery or university, locked up in chests and cupboards. This in turn affected the arrangement of the Fellows' living quarters, because unlike a monastery the college had no system of separate carrels. And just as our lifestyles today are so influenced by what we take from our college days, so the arrangement of what was effectively a college dormitory room in the fourteenth century in turn affected the way books were kept and used subsequently in private homes.

During the Renaissance it became increasingly popular to have a personal study, either in the corner of one's bedroom or in a small but separate room. Such studies were generally crowded but secure places, ideally located in quiet and remote areas of the house; when

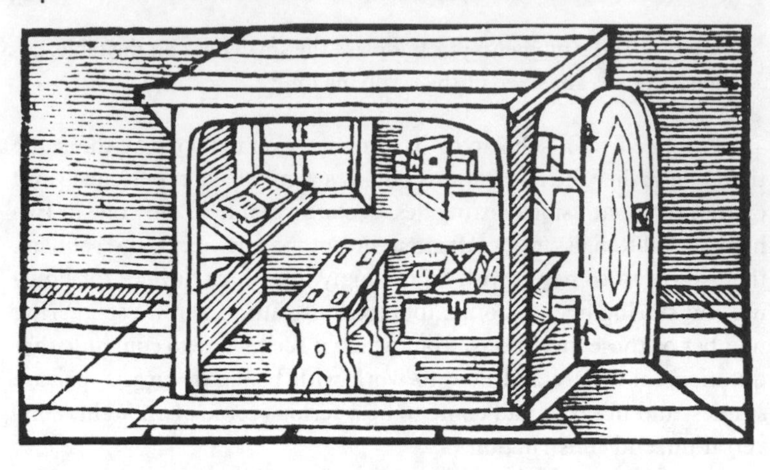

This enclosed study, depicted in a 1539 woodcut, appears to have been built in a larger room to secure a private space. The cutaway wall shows that the study was located at a window, even at the expense of obstructing light to the larger space. The top of the book chest serves as a table on which to lay books flat, and the shelf above it holds books leaning against the wall with their covers facing out.

locks and keys were considered necessary, the locks might be installed on the door to the larger chamber and not on the door to the study proper, which might only have occupied an alcove that opened into the room or have been situated, perhaps on a raised platform, in a corner or beside a window. Locating the study near a window as in New College was, of course, necessary; the presence of daylight by which to read and work was of the utmost importance. Just as the number and location of windows in institutional libraries, whether in monasteries, churches, or universities, would increasingly become a determining factor in the configuration of carrels, libraries, and bookcases, so windows in studies very much controlled the configuration of furniture.

In the private study, books were, of course, unchained and likely to be left out in the open. Those that were kept in chests, which may or may not have been fitted with locks, tended to be the less frequently used or more valuable items in the owner's library, and the chest may have kept the books free of dust as well as, for security reasons, out of

This woodcut from the 1517 calligraphic treatise of Francesco Torniello shows him working at a trestle table in a tight space beside a window. Note the arrangement of the books faced cover out on the narrow and lipped shelf in the background.

the way. Books that were frequently consulted came increasingly to be placed on shelves, which typically rested on brackets anchored in the wall. A shelf attached to a wall was thus a common fixture in the study, a convenient place to keep not only books but also a store of ink and other writing paraphernalia off the desk and out of the way.

There seems to have been nothing of the item that we would call a bookend, and only in rare cases does it appear that books themselves were used as one, as they sometimes are today, by being piled up horizontally so as to support other books in a vertical position. Depictions of books in vertical positions are so uncommon, in fact, that one has to assume they were accidents of what often appear to be random or haphazard arrangements or evidence of artistic license—to make a well-composed still life of fronts and edges of books. The nature of studies and the books in them is very evident in the many renderings of St. Jerome and other scholars that exist in paintings, illuminated manuscripts, and early printed works that show nooks and crannies where books are stashed—in cabinets beneath the desks, on shelves in front of them, in the triangular attic-like spaces formed under the

This late-fifteenth-century bibliomaniac in his study is remarkable for wearing a pair of eyeglasses, but note how his books are arranged every which way on shelves and in the cabinet beneath the double-sided lectern-desk.

back-to-back sloped surfaces of desktops or small tabletop lecterns that rested upon a horizontal surface. The larger and more elaborately decorated books continued to be displayed face out on nearby tables and on the shelves—slanted and horizontal alike—mounted on the walls of the scholar's study; but as more and more books became acquired, where to store them was a growing problem. The custom of displaying books leaning against the wall on flat shelves or face up on lectern-like shelves was not given up easily, and so when there was no

A Carmelite monk is shown here working in his study at a desk mounted on a swivel, which would enable it to be moved out of the way. Note the books neatly arranged face out on the sloped shelves, in front of which curtains could be drawn to keep out light and dust.

more room on the tables and shelves, the overflow volumes were placed everywhere one could find space.

St. Jerome in his study was a common subject of Italian and other European artists in the fourteenth century, and it was a theme returned to frequently by Albrecht Dürer, the foremost German engraver and woodcut designer of the early sixteenth century. As a young journeyman artist in the late fifteenth century, Dürer traveled rather extensively and would no doubt have heard about and even seen a good number of the many treatments of St. Jerome that had been rendered in paintings and in book illustrations. It is not remarkable, therefore, that Dürer's engravings and woodcuts of the saint include many of the same accouterments that other artists presented in depicting Jerome's most celebrated activity—writing. Indeed, Jerome was famous for producing, among other significant works, the Latin translation of the Bible known as the Vulgate, so called because it was rendered in the common or "vulgar" Latin of the time and so was more generally accessible than the original Hebrew and Greek in which the Bible was written.

Jerome, whose name in Latin was Eusebius Hieronymus, was born in the middle of the fourth century in what later became Yugoslavia. He studied in Rome and subsequently lived in a wide variety of settings, including as a hermit in the desert. It was there that he reputedly once helped remove a thorn from the paw of a lion in distress—hence the seemingly incongruous appearance of contented lions in so many depictions of Jerome in his study. The saint wrote a considerable number of works on ecclesiastical history and biblical exegesis, and as a Doctor of the Church, had an enormous influence on scholarship in the Middle Ages. It is not surprising that he was the subject of so many paintings, engravings, and woodcuts. The degree of verisimilitude in these portrayals is open to debate, however. As we have seen, in some cases Jerome is writing on a scroll, something he is unlikely to have done, the codex having replaced the scroll for Christian texts. Indeed, it was Jerome himself who recorded in the fourth century that damaged papyrus scrolls were being replaced with copies on vellum in the Library of Pamphilus at Caesarea, which was on the Mediterranean coast of what would become Israel. What appears to be most likely is that the artist, painting a full millennium after his subject lived, furnished the study with imagined activities of old and trappings contemporary not with Jerome but with the more familiar fourteenth and fifteenth centuries, a reasonable exercise of artistic license. Whether the visual clues about the care of books are early Christian or Renaissance, however, they are different from what we know today.

One fifteenth-century representation of Jerome at work is in an oil painting by Benedetto Bonfigli, which hangs in the church of St. Peter in Perugia, in central Italy. Here the saint sits before a freestanding desk of a kind frequently found in contemporary paintings. It has a sloped top, and in its base is a cabinet whose door is opened to reveal books stored inside. St. Jerome appears to be consulting one of the

St. Jerome is using a rotating book wheel, which appears to be capable of holding four open books, in this fifteenth-century oil painting by Benedetto Bonfigli.

several codices propped open on what looks to be a revolving lectern-like device, a common piece of furniture appearing in medieval depictions of scholars' studies. The mode of composition of books in Jerome's time, and to a great extent through the Middle Ages, was to compare many texts and to copy freely from them. Hence, what today might be described as a lazy Susan of a book stand was an extremely handy thing to have in one's study. There is a tapestry behind Jerome, which may hide books on bookshelves (a curtain was often draped in front of books to keep out light and dust). The books visible inside the cabinet are neatly arranged, but they are horizontal on the shelves, piled one upon another, and they have been placed into the narrow cabinet top edge first, so that what one sees are the bottom edges and fore-edges of the books, neatly clasped. There is one scroll rolled up in the corner of the cabinet, but there is no indication that it is being consulted.

In Dürer's 1492 woodcut *St. Jerome Curing the Lion,* books are

St. Jerome Curing the Lion, *a 1492 woodcut by Albrecht Dürer, shows some of the scholar's books opened for study, with others arranged haphazardly on the small high shelf.*

open on a variety of lecterns, showing Greek, Hebrew, and Latin texts. A cabinet is built into the base of the desk, and the open door allows us to see that it contains not books but what appears to be a flask, probably holding ink, and a variety of other less distinct objects. There are a few books above the desk, some of which are clearly on a shelf mounted high on the wall. The shelf appears to be supported by brackets at either end, a common means of shelf construction at the time Dürer was working. Such shelves are frequently present in the background of paintings and drawings, and they seem to be of the type we today would recognize as similar to the adjustable bookshelves that are supported by brackets that fit into slotted strips of metal anchored to the wall. In Jerome's, and still in Dürer's time, shelf brackets might even have been mounted directly into a wall during its construction. Any such projection from the supporting wall is known as a cantilever, the structural mechanics of which Galileo would explore in a seminal way in his 1638 treatise *Dialogues Concerning Two New Sciences.*

By the placement of a board across two cantilevers or brackets located at the same level, a shelf is formed. Our main interest presently in the shelf Dürer incised into wood is, however, not the brackets or even the shelf itself but the books upon it. There appear to be three of them, but they seem to be haphazardly arranged. They are certainly not standing upright. Nonetheless, even though *St. Jerome Curing the Lion* has been described as "somewhat ungainly and archaic," there is no reason to believe that it did not portray accurately in some sense what Dürer saw in his visits to the studies of fifteenth-century scholars. The lion does seem poorly executed, and the woodcut's perspective and other features certainly do not show the artist at his mature best, but the locus does appear to be a study to be studied, if we are so inclined. In particular, the way Dürer arranged the books on the shelf cannot have been too far from the way books were found in active studies, though we must allow that they might have been arranged here for their compositional value, much as a still-life painter might arrange fruit in a bowl or flowers in a vase. However, if it was common at the time for books to be standing upright between bookends, as they might be on a shelf today, it would seem that Dürer would certainly have so rendered them.

When Dürer returned to the subject almost two decades later, his technique had improved greatly, and in his 1511 woodcut *St. Jerome in His Cell* we see a more competently rendered lion, saint, and study. Things generally are more orderly, and the details are considerably refined. Behind Jerome is a long, high shelf which holds such things as a candlestick and flasks, the latter again presumably used for storing ink. On a lower level, beside the working saint, is a more easily accessible shelf with books upon it. The books are all closed, but none of them is in a modern upright position. One book lies on its back, with its bottom facing out. Three books of unequal width are sitting on their fore-edges, with their bottoms out, as we might find books arranged today when they are offered at bargain prices out in front of

Dürer's 1511 woodcut of St. Jerome in His Cell *shows books on a shelf near at hand for the scholar's use but not arranged in any single way. The book on the chest in the foreground has a bookmark placed near the fore-edge, a practice that could be followed because the pages were held tightly together by the clasps.*

a used-book store. On top of these books lie two more, neither of which has its spine facing out. This bookshelf is the only item in Jerome's study that appears by modern standards to be in disarray.

In Dürer's most famous treatment of the subject, his 1514 engraving *St. Jerome in His Study,* there is again a high shelf holding candlestick and flasks, but the books now sit not on a wall shelf but on the windowsill and on a window seat before it. The books are arranged square with the sill, but none of the four shown is in the spine-vertical

Dürer's 1514 engraving of St. Jerome in His Study *shows books neatly arranged on the window seat cum shelf, but with neither fore-edge nor spine facing outward.*

position. Three of the books are resting on their fore-edges, with their spines up, and a fourth, larger book is resting on its back, with its top facing upward. Books placed with their fore-edges down might readily stand in place, which they certainly would not have done if put down on their rounded spines, for they would have rolled to one side or the other. But it is also important to note that if some of the spines were facing up, it was not necessarily for identification, for no identifying marks were typically imprinted on a book's spine in Dürer's time.

In all of these examples, the explanation for the general lack of an orderly arrangement of books of the kind we would expect today is on the one hand straightforward. The depiction of the scholars was of them at work, when their books were likely to be in disarray, and not neatly piled up or arranged as they might be after one writing project was completed and before another was begun. Scholars were not apt to own a great number of books, and they would borrow what they needed for a project and return the books after they were finished with them. The private ownership and arrangement of books as we know them were certainly not common in Jerome's time, nor even yet extensively so in the fifteenth century when the saint was depicted by Dürer.

There were, however, a growing number of eccentrics whose book-collecting and -storage had begun to interfere with their lives. Richard de Bury, a fourteenth-century bibliophile, was "compelled to climb over his books to reach his bed." Indeed, after the printed book all but replaced the manuscript in private libraries, the problem of reaching one's bed could become even impossible. Thus in the eighteenth century, Thomas Rawlinson—"who gathered books much as a squirrel gathers nuts," who was alleged to have read title pages and little else, and whom Joseph Addison nicknamed "Tom Folio"—had so many books stuffed in his rooms at Gray's Inn that he had to sleep in the hallway.

Between the extremes of the careful scholar or monk who read and reread the few books that he had in his carrel and the collector who looked for every which way to fit more and more titles into his storage place—such as the eighteenth-century "bibliocast and shoemaker, John Bagford, who collected not books but title-pages," which in time filled sixty-four folio volumes—there were those whose large but necessary libraries needed new solutions for the problems of book storage and use. Among the devisers and inventors of library furniture was the sixteenth-century Italian military engineer Agostino Ramelli, whose *Diverse and Ingenious Machines* was published in 1588. This book, which falls into the genre of illustrated printed works known as "theaters of machines," is filled with almost two hundred 6-by-9-inch engravings of everything from grain mills to siege engines. Unlike many of Leonardo's notebook sketches which leave much to the imagination, Ramelli's drawings are extremely detailed and well developed.

Among the mental constructions that Ramelli describes is a revolving desk resembling a water wheel which is like nothing known to have been seen in any contemporary Western study. Indeed, Joseph Needham, the scholar of Chinese science and technology, has argued

Agostino Ramelli's fanciful book wheel was illustrated in his 1588 theater of machines.

that a revolving bookcase had its origin not in the West but in China, "perhaps a thousand years before Ramelli's design was taken there." According to Needham, "the fact that Ramelli's was a vertical type, and that all the Chinese ones, from Fu Hsi onwards, were horizontal, would simply have been characteristic of the two engineering traditions," illustrating "perfectly the preference of Western engineers for vertical, and of Chinese engineers for horizontal mountings." Whether this be a valid generalization may be argued, as may Needham's further speculation that "probably from the beginning, however, the rotation was a piece of religious symbolism as much as a convenience."

There are, of course, many horizontally mounted rotating bookcases depicted in Western art, not to mention realized in Western manufacture, and this kind of inventiveness was not to end with the Renaissance. A Victorian guide to the home library noted that "A kind of square revolving bookcase, an American invention, manufactured by Messrs. Trubner, is useful to the working man of letters. Made in oak, stained green, it is not unsightly."

Whether such devices were appreciated most on grounds of aesthetics, symbolism, or scholarly convenience may have to remain a matter for speculation. There can be little doubt, however, that many a scholar who used such rotary devices in the course of copying, translating, and explicating found them a godsend. The Ramelli wheel may or may not have been so practical, however, for while the illustration of it shows a reader able to consult a series of books as we might click backward and forward from web page to web page on the Internet today, there does not appear to be any convenient working surface on or near the wheel for the scholar who may wish to make notes or write. If a further anachronism may be allowed, the device looks like a 7- or 8-foot-tall model of a Ferris wheel, with open books riding on individual lectern cars, suited for passive or recreational reading but not active scholarship involving writing. Nevertheless, according to Ramelli,

> This is a beautiful and ingenious machine, very useful and convenient for anyone who takes pleasure in study, especially for those who are indisposed and tormented by gout. For with this machine a man can see and turn through a large number of books without moving from one spot. Moreover, it has another fine convenience in that it occupies very little space in the place where it is set, as anyone of intelligence can clearly see from the drawing.

The capacity of the wheel appears to be a dozen books, and the reader sitting before it seems to be turning it with his hands, which can conveniently grasp the large, solid-looking side wheels. In the tradition of Agricola—whose earlier sixteenth-century work on mining is illustrated with much machinery, with exploded views employed to show details of construction otherwise hidden—Ramelli cut away some of the wheel to show its hollow interior, in which an arrangement of planetary gears engaged each other in such a way that the lecterns would not swing freely like the cars on a carnival ride but would be held at the same angle to the floor no matter where in their passage they happened to be. This feature of the revolving desk was essential, of course, lest in the device's use the books would be tipped and fall off their perches. Ramelli's further description of the book wheel confirms that the advantages of the device were in what it did not do and did not require:

> This wheel is . . . constructed so that when the books are laid on its lecterns they never fall or move from the place where they are laid even when the wheel is turned and revolved all the way around. Indeed, they will always remain in the same position and will be displayed to the reader in the same way as they were laid on their small lecterns, without any need to tie or hold them with anything.

The attention to detail that Ramelli showed in the design and operation of his revolving reading desk provides strong evidence that other details in his illustration should be taken seriously. The door in the background, for example, is fitted with a lock and two sliding bolts, an arrangement not unlike what we might find today on the door of an apartment in a large city. The level of detail is such that the bolts can be seen clearly to be slid to the open position. We can assume that they would be in the locked position if the scholar wanted to ensure his own security or that of his books, or merely his privacy, when he was absorbed in reading or was afraid that he might doze off at his wheel or fall asleep in the bed that very well might have been located out of the picture's frame or behind the observer.

Two other details in the illustration are worthy of note. Next to the door there is a case of three bookshelves, of a design that evolved from the lectern system when chains were no longer required. This is another chapter in the story of the bookshelf, as we have seen, but we can note here that Ramelli's case has an interesting arrangement of shelves in that the lowest is about 3 feet off the floor, and the space beneath it is not used for books. The shelves reach to the ceiling of the room, which must be 8 or 9 feet high. Getting a book from the top shelf would involve a stretch that could have been avoided by having put the third shelf below the bottom one. As was the case in institutional libraries, however, the arrangement depicted by Ramelli was the way bookcases did in fact evolve—upward toward the ceiling before they reached down to the floor.

Another interesting thing about the bookcase in Ramelli's drawing is its location by the door, against the wall. The shelves are actually crowding the door and, given the way the door opens, would have been a hindrance to easy ingress and egress from the room. Had the shelves been constructed on the outside wall, which appears to be bare in the engraving, access to the room would have been much more convenient. However, to have placed the bookcase there would have put the books in the shadow of the wall in which the window was located. Thus, given the constraints of the room and its use, Ramelli positioned the bookcase where the optimum amount of light would reach the books and the reader consulting them, even though they were shelved a good distance from the window.

The preferred location for such a case would be perpendicular to and much closer to the window, where the light was brightest, but in Ramelli's room arrangement the revolving desk occupies that space. And the reason the desk is beside the window is because having light on the books being read is the first priority. In general, the depiction of the studies of scholars in fifteenth-century illuminated manuscripts shows the careful thought that must have gone into locating desks relative to windows, the source of natural light, which was much the preferred way of illumination for reading and writing, for candles and oil lamps were not only hard on the eyes but also presented a clear danger to the books of the scholar who might fall asleep over them. Where this might be a danger, the Chinese story of the "ideal scholar" could be held up as a model, "for he was in the habit of fastening his queue to a beam, so that if, when overcome by drowsiness, his head fell forward, he would be awakened by the pull on his hair."

One last detail in Ramelli's illustration calls out for attention, and that is the fact that the books on the shelves are all arranged vertically, with their spines out. In this regard, Ramelli was as forward-looking as he was inventive. There appear to be few if any other contemporary depictions of books on shelves that have the same rigid arrangement that Ramelli shows. In the sixteenth century it was still much more common for books to be displayed on a sloping shelf with the decorated front exposed, as they were on lecterns in monasteries, to be leaned against the wall behind a shelf, or to be placed flat on a horizontal shelf with their top, bottom, or fore-edge out. When books were shelved in a vertical position, it was the fore-edge and not the spine that was faced out, as they were in the library of the dean of Canterbury, John Boys, who in the early seventeenth century still "clung to ancient fashions so far as to set his books with their fore-edges outwards," even though they were arranged on shelves "of a modern type" as in Ramelli's illustration. In fact, the turning around of books on the shelf followed the introduction to institutional libraries of bookcases whose impact on the care and use of books was even more revolutionary than that of Ramelli's fanciful wheel.

Private libraries, because they did not contain nearly as many books, and because those they did house were not chained, were generally not under the same constraints of space as were the older institutional libraries. Smaller libraries thus could develop different solutions to the problems of light and space. Indeed, "before the seventeenth century, the normal English private collection rarely amounted to more than a few score volumes, probably stored in an oak chest, or laid flat on a table, or possibly kept on a shelf or two fastened to a wall." There were exceptions, of course. The one or few bookcases that were needed to house a larger private collection were accommodated in rooms that might or might not have been fitted with ideal arrangements of windows.

In any case, as with those of St. Jerome, Renaissance depictions of scholars' studies seldom show books arranged in a such a way that all of them are oriented the same. Typically, some books would be shown with their fore-edges out, others would have the top or bottom out, and still others would be leaning against a wall or a horizontal pile of books showing off a decorated front cover. This haphazard arrangement is how one would expect books to be found around a desk and on shelves of a working scholar. Because there typically would not be that many books in a private library, the scholar could be expected to know each of his books by size and thickness, by color and texture of binding. Thus there was no need to mark books or arrange them in any systematic way, because any of a score or so of volumes could be located in an instant.

In larger libraries, titles were sometimes written on the fore-edges or top or bottom, as modern schoolboys have written the name of a book's subject on the edges of its pages. At least one sixteenth-

The library of John Boys, dean of Canterbury Cathedral, in a sketch published in 1622, is shown fitted with modern wall shelving but retaining the practice of arranging books with fore-edge outward and their clasps showing.

century Italian book collector, Odorico Pillone, had the artist Cesare Vecellio paint the fore-edges of his books with scenes appropriate to their contents. In all, 172 books were so decorated, including two that were painted upside down, which one scholar has interpreted to mean that artists were not totally accustomed to painting books in this way because it was not a common practice. The decorated fore-edges of Pillone's books were also lettered over with literary identifiers, further suggesting that the purpose in part was to identify individual books in this large library, some of which at least were indistinguishable by their size and binding alone. Three volumes of the works of St. Jerome bound in wooden boards had their clasps removed, apparently so as not to obstruct the fore-edge paintings, which depict the scholar in his study and in the desert.

Prior to the introduction of printing from moveable type, many books were bound in elaborate and unique ways, often with their boards covered in leather or fabric and sometimes overlaid with metal bosses, carvings, and jewels. Many of these latter "treasure bindings" were so valuable that during "the spoliation of the monasteries under Henry VIII, and the wholesale destruction under Edward VI, of all vestiges of the old learning," orders were given "to strip off and pay into the king's treasury all gold and silver found on Popish books of devotion."

In New York City's Pierpont Morgan Library, a repository of medieval and Renaissance books in treasure and fine leather bindings, there are many wonderful examples of the bookbinding art. The beautifully illustrated book *Twelve Centuries of Bookbindings, 400–1600* pictures and describes some of the Morgan collection, more or less in chronological order; it is instructive to page through the work—from front to back—to see how the decorated spine evolved. It is almost as striking as the depiction of the evolution of a bent-over ape into a man standing erect. The earliest books pictured have their leather-sheathed wooden covers heavily bossed and jewel-encrusted, with the three dimensional nature of their covers making it all but impossible to shelve them in any way but on their back covers, which were also often heavily ornamented but in a more planar fashion. Such books could not safely be stood upright the way modern volumes are shelved.

The spines of early books had been modest features indeed compared to the front and back covers. (Charles Dickens would later write in *Oliver Twist* that "There are books of which the backs and covers are by far the best part.") In many cases, the metal and jewel treatments of the front were nailed or otherwise fastened directly over leather or other binding material, emphasizing the plainness and subservient position of the spine. There was little that could be done otherwise, for the spine was in effect the hinge of the book, something that had to bend and flex if the book were to open properly, and so it was not suitable for heavy ornamentation. Indeed, the spine was to the cover as the downstairs would be to the upstairs in a Victorian mansion.

As elaborately tooled leather bindings came to be more fashionable than repoussé and other three-dimensional treatments, it became increasingly possible (not to mention necessary in increasingly crowded libraries) not only to shelve books vertically front cover to back cover, in the modern mode, but also to decorate the spine to a degree commensurate with the front and back. After all, typically the

This fifteenth-century German codex has its edges decorated, a practice that helped identify books shelved with their bottom or fore-edge out.

same piece of leather encased the entire book. The exoskeletal spine, which holds up the innards of the book structurally in a fashion not unlike the way our own spines hold us up, was still the machinery of the book, however, and so it continued to be the part that was hidden as much as possible, pushed into the dark recesses of bookshelves, out of sight. Shelving books with their spines inward must have seemed as natural and appropriate a thing to do as to put the winding machinery of a clock toward the wall or behind a door, or both.

As long as there were relatively few books in a library, they could be located, with or without a list of a shelf's contents, and be identified without being inscribed. Which book was in each distinctive binding would be known as we are aware what grains are in the unmarked canisters on our kitchen counter or what is in the decorative boxes in which we keep odds and ends. We have a pin box, a button box, a coin box, each identified by the individual container's distinctive shape, size, and color. When we begin to accumulate too many different canisters or boxes, however, we are likely to find ourselves going to the wrong box more often than not. It is at that point that we tend to begin to label them. A similar problem occurs when we have a lot of different things in similar-sized and -looking containers. It is the rare spice rack, for example, that will not have its identical little jars or tins labeled with the name of the spice contained.

So it was with books, as they became increasingly numerous and similar-looking, especially with the introduction of the fashionable practice of binding all the books in one's library to match. Indeed, it is rare that a book produced before the fifteenth century has not been rebound, and so few original bindings from that period survive. In time, in the course of rebinding old volumes and the general adoption of marking spines with some identification of the book's contents, all books came to be similar in the way they were bound and shelved with their spines out. Such an evolution in bookbinding and, hence, book shelving, took on the order of twelve centuries, however, and from that perspective it is not difficult to see that there remains confusion in the language as to whether the spine, or hinge, of a book is its facing or backing.

Some private owners came to be very particular as to how their books looked and how they shelved them. Beginning in the mid-sixteenth century, the decoration of the spine began to be "brought into harmony with that of the sides." With the introduction of uniform leather bindings that were continuous from the front to the back cover, the spine of the book increasingly was decorated not only with designs in harmony with the rest of the binding but also with the name of a book's author or title and date of edition.

Not surprisingly for a new practice, whether the title should read up, down, or across the spine was not agreed upon. Indeed, lack of agreement in the matter persisted among English-speaking countries as late as the middle of the twentieth century, when books bound in Britain still tended to have their titles read up the spine, whereas those from America read down, conventions as opposite as driving on different sides of the road. In time, the British way of book labeling gave way to the American, which can be argued to make more sense because when the book is lying face up, the title may be read easily. Of course, in the old British way a book's spine could most easily be read when the book was lying face down, in which case any identification on the cover—increasingly the dust jacket—could not be read. Uniform practices still do not exist in non-English-speaking countries.

The addition of an author's name or the title of a work to the spine of a book was evidently done in France, as well as in Italy, before 1600, providing strong evidence that these books were shelved spine out. In fact, of all old books, "the earliest tooled in gilt on the back seem to have been done in Venice or northern Italy about 1535." That is not to say that all books in a library would yet have been shelved with spines out, as demonstrated by the fore-edge-painted books of the Pillone library, which date from about 1580. However, through that time at least it appears to have remained the custom in other countries, such as Germany, Holland, Spain, and England, to shelve books spine in, "following the fashion of the chained library, where this has to be done for mechanical reasons," as we have seen. (Custom may also affect how books are shelved: the books in the library of the Spanish Escorial were still shelved with their fore-edges out in the late twentieth century.) This was by no means a universal practice in the sixteenth century, however, and anyway there was the fact that some books would be identified on the spine while older ones would be marked on the fore-edge. Such a condition naturally led to books being shelved some in the old way with their fore-edge out and others in the new way with their inscribed spines out.

No doubt the reality is that there was a period, which might have spanned the greater part of a century or more, during which older books were shelved spine in and newer ones, or rebound ones, spine out. They might have been segregated in different cabinets or presses according to this scheme, or they might have been inter-shelved. When we grow up in the midst of a changing technology we tend to be very tolerant of anachronistic juxtapositionings—as so many of us know who work at computers set up on an old-fashioned desk whose surface is too high for typing comfort, but to which we have adapted.

One of the first large libraries to be arranged with all book spines outward was that of the French politician-historian Jacques-Auguste de Thou, who is noted for being a pioneer in the scientific approach to history. He had one of the most impressive libraries of the late sixteenth and early seventeenth centuries, and at about eight thousand volumes the collection was also so large that how de Thou arranged it was naturally important to his use of the library, and of interest to those who knew of it. Although Ramelli was Italian, he was in service to the king of France when his theater of machines was published, and so the arrangement of books spine outward in the background of his 1588 etching of a book wheel depicts a practice that appears to have been evolving at least in France toward the end of the sixteenth

In the sixteenth century, books began to have authors and titles, and the date of the edition, imprinted on their spines. As long as the practice was far from universal, however, not all books were shelved spine outward. Here, a book not so imprinted is identified by a slip of paper tipped into an inside cover and folded over the book's fore-edge.

century. In the seventeenth century, it became increasingly the case that the spines of virtually all books were decorated and lettered, something that had generally not been done before.

One curious binding practice of sixteenth-century Germany consisted of two books bound back-to-back sharing the "cover" separating them and thus using only three boards instead of four. In order that the back-to-back books would each open in the conventional way, they would have to be bound in opposite directions, with the one's spine juxtaposed with the other's fore-edge. These "dos à dos" bindings, as they have come to be called, were not common, but they suggest an accommodation during the transition period when it would not have been unusual to find both spines and fore-edges facing outward on the same bookshelf.

The diarist Samuel Pepys had one of the most extensive libraries in seventeenth-century England, and in 1666 he had it refitted with new bookcases, when "his books were growing numerous and lying one upon the other." At first he appears to have had only two bookcases, for within a year he wrote in his diary:

> The truth is, I have bought a great many books lately to a great value; but I think to buy no more till Christmas next, and those that I have will so fill my two presses, that I must be forced to give away some, or make room for them, it being my design to have no more at any time for my proper library than to fill them.

In spite of his intentions Pepys, like many a book buyer since, eventually bought more presses to hold his books. The first cases were made for Pepys (who would later become secretary of the Admiralty and still later president of the Royal Society) by Thomas Simpson, a dockyards master joiner. The cases of carved pickled oak show signs of nautical construction, with their broad base reminiscent of a book chest but with doors that slide up and down rather than swing on hinges. Such features would not be practical for the tops of the cases, but they do have sturdy sliding locks that would keep their doors from swinging open in even the roughest seas. Twelve cases in all are preserved today in Cambridge's Magdalene College, where they hold about three thousand books of the Pepysian library as it was, in accordance with Pepys's will, put in final order by his nephew. Not all of the book presses are identical, for they were not all made at the same time or by the same hand, but they appear to be identical upon first glance, and they make for a strikingly fitted-out library indeed.

Pepys limited his library to three thousand volumes, numbered from 1, the smallest in size, to 3000, the largest. When he acquired more books than he could shelve in style, Pepys discarded less-wanted ones to make room for the new. In time, he found that there were ways to find more space for books even on a limited number of bookshelves.

To conserve room in his "closet," as a private study was then called, Pepys arranged his books in double rows, with a slightly raised narrower shelf holding taller books above and behind a line of smaller ones on the main shelf. Such raised shelves also equalize the gap between book tops and the shelf above.

The strict arrangement of Pepys's books by size, "the placing as to heighth," is visually very striking, with the smallest books being held on the lower shelves of all the bookcases, and with the size order continuing around the library room (an arrangement no doubt deplorable to the row-oriented Melvil Dewey) with an almost imperceptible increase in book height. The front edge of each wide adjustable shelf is at such a height as to be in line with the horizontal framing of the panes of glass in the bookcases—among the first to be so glazed. Replicas of the Pepysian bookcases can be ordered, but the example displayed on the brochure advertising them shows the shelves placed with no regard for the frames of the panes of glass and the books arranged in some scheme other than size. The effect is to make both the books and the bookcase appear awkward and unkempt, thus diminishing the appeal of each but emphasizing the thoughtful order of Pepys's cases.

By choosing to arrange his books by size, Pepys achieved a striking appearance that would have been lost by a grouping according to subject matter or any other scheme. Where an otherwise matched set of books contained volumes of different sizes, perhaps because they were printed at different times in various formats, Pepys had blocks of wood carved and decorated to match the shorter volumes and lift them up so that they matched in height their mates. Where books were printed in landscape mode, Pepys shelved them on their foreedge, so as not to have them project too far front or back and thus interfere with the arrangement he devised.

By Pepys's time the custom of identifying books by imprinted spines had been well established, and so virtually all of his books were shelved spine out, and he could see clearly the titles of these books, even those in the back row, because they were sufficiently higher than those in the front. Pepys was proud of his arrangement of bookcases in his new study, and he wrote in his diary, "I think it will be as noble a closet as any man hath, and light enough—though indeed it would be better to have a little more light." As good as it was, there was always

One of Samuel Pepys's book presses in the Pepys Library at Magdalene College, Cambridge, is shown here preserved in the manner that the diarist desired. The books are arranged in order according to size, with the shortest books occupying the bottommost shelf of the main section in all twelve bookcases. The depth of the cases is such that, except for the shelves holding the tallest books, a second shelf holds another row of books behind the front one.

room for improvement. In 1680 Pepys added an elegant writing desk, the "earliest known oak pedestal writing desk," with space at its sides to shelve behind glass those books too tall to fit in his orderly bookcases. The idea of putting a freestanding desk or table in a room with bookcases placed flat against the walls was not yet a common idea, but it would become increasingly so in time as libraries of all sorts added furniture unrelated to the function of shelving books.

CHAPTER SEVEN

Up Against the Wall

The stall system—the arrangement of fixed bookcases perpendicular to exterior walls pierced by closely spaced windows—was long a characteristic of English institutional libraries. On the Continent, a different tradition evolved independently, in which the bookcases were arranged parallel to and against the walls, so that a visitor entering such a library saw not row upon row of bookcases forming carrel- or alcove-like recesses in front of each window, but rather a room encircled with bookcases all facing a great open space in the center. This wall system, as it has come to be known, is believed to have been first introduced on a large scale in Spain's El Escorial.

The Escorial was ordered by Philip II to be built near Madrid as a monastery and royal pantheon, where most Spanish sovereigns would subsequently be buried. The massive structure, begun in 1563 and completed in 1584, housed the royal apartments and a great library. The room occupied by the library, measuring 212 by 35 feet, is characteristically long and narrow, albeit exceptionally long, but there ends its resemblance to English libraries furnished according to a lectern or stall system. The Escorial library is topped by a 36-foot-high barrel vault, and the east and west walls are pierced not by numerous narrow and closely spaced windows but by a dozen large windows, reaching 13 feet up from the floor, which makes the room very well illuminated by natural light. The wall spaces between these windows are occupied by the bookcases, which are heavily ornamented with fluted Doric columns and heavy cornices. The bookcases are fitted with desks, and, according to John Willis Clark, who wrote at the turn of the twentieth century,

> The desks are 2 ft. 7 in. from the floor, a height which corresponds with that of an ordinary table, and suggests that they must have been intended for the use of seated readers, though seats are not provided in the library at present. The section of the shelf and desk placed beside the elevation shews that there is a convenient slope to lay the books against. The uppermost of the four shelves is at a height of 9 ft. from the ground, so that a ladder is required to reach the books [which] have the fore-edge turned outwards, according to what is, I am informed, the usual custom in Spain.

Although the Escorial bookcases might be positioned differently from their contemporaries in England, they clearly have many of the same features. In particular, they are fitted with a desk; this suggests that the books were originally chained to the presses, as does the custom of shelving the books spine inward. The Escorial desk has the thoughtful feature of a slanted surface against which to lean books. The presence of large windows, which let in plenty of light, some of it reflected off the vaulted ceiling, evidently meant that it was possible to face the bookcase or sit at the desk attached to it and still have enough light by which to read. According to Clark, "the Escorial had a very definite effect on library-fittings elsewhere, but like other important inventions, the scheme of setting shelves against a wall instead of at right angles to it occurred to more than one person at about the same time," and so he could not "construct a genealogical tree."

In the early seventeenth century, the important Biblioteca Ambrosiana was built in Milan, Italy, and it took the wall system a definite step higher. The 74-by-29-foot room had, like the Escorial, a barrel-vaulted ceiling, but instead of windows in its long walls it had at each end of the vault a great semicircular window, which let in plenty of light to illuminate the room and its books. Because there were no windows in the lower wall, it was completely lined with bookshelves, save where there were doorways to adjacent rooms and spaces. These bookcases reached up to 13 feet above the floor, and there was an additional band of shelves 8 feet 6 inches tall over these. They were reached via a gallery, which wrapped around the entire room. Access to the 2-foot-6-inch-wide gallery was by staircases in each of the four corners of the room.

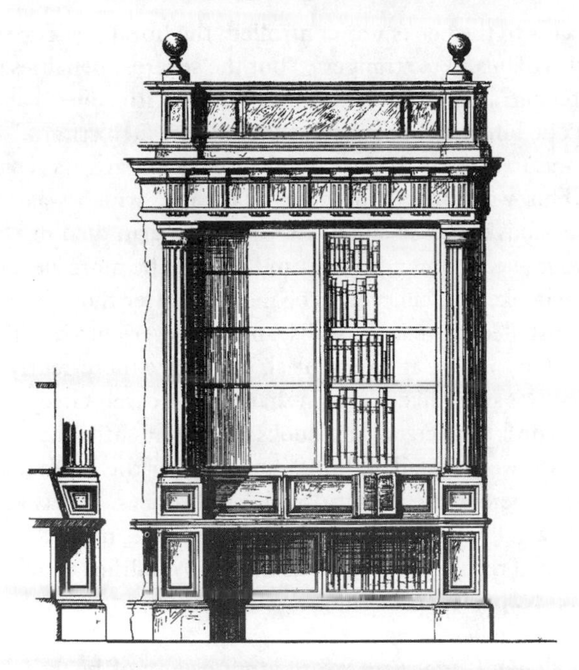

Bookcases installed against the wall of the library in the Escorial had a desk integrated into their design, an indication of the fact that the books were originally chained to the shelves. The open book shown here is resting against the sloped back of the desk.

There were no desks at the bookcases, as noted by a contemporary writer, who remarked that "the room is not blocked with desks to which the books are tied with iron chains after the fashion of the libraries which are common in monasteries, but it is surrounded with lofty shelves on which the books are sorted according to size." While the books were not chained, as late as the early twentieth century they were "protected by wire-work of an unusually large mesh, said to be original." The books on the higher shelves and those above on the gallery did not need to be so protected, for access to them was only by ladder or via the stairways. In other words, no book could be reached to be taken off a shelf without permission, and only then could it be consulted or read at one of the tables in the corners of the room. Even though access to the books was controlled, the library was open to the citizens of Milan and to strangers, "but the severest penalties awaited those who stole a volume, or even touched it with soiled hands; and only the Pope himself could absolve them from such crimes."

What was to become the first public library in Paris was completed in 1647. This was the Bibliothèque Mazarine, which was unfortunately built above stables. When Cardinal Mazarin died in 1661, the library was moved to finer quarters and came to be more elegantly situated. The bookcases, believed to be modeled after those in the Escorial, were installed against the walls, and they were fitted with a desk somewhat akin to those at the Escorial, but with a more gently sloping surface that was more like a lectern than a book easel. Given the presence of the desk or lectern, the books were evidently chained when the bookcases were installed, but later those on the lower shelves at least came to be enclosed behind wire mesh doors. It was agreed in 1739 to add a gallery, for additional shelf space was needed. Since the roof was in bad repair, the vaulted ceiling was modified to a flat one so that shelves capable of holding twenty thousand books could be installed in the gallery.

The Bodleian Library at Oxford is a museum of bookshelves, the oldest of which are arranged according to the stall system, but the earliest additions to which would be constructed according to the wall system. The central core portion of the library dates from the 1480s, when it was moved into the then-new Divinity School Building. The Divinity School itself occupied the ground floor, and the library was located on the first floor above the ground floor, what in America is called the second floor. Having the library a floor up restricted access through well-controlled stairways and reduced considerably the possibility of unauthorized access through the windows. This last was a not insignificant concern, for protecting the chained and unchained books alike was a foremost responsibility of the librarian.

The central library at Oxford was known as Duke Humphrey's Library because it was Humphrey, Duke of Gloucester (a brother of Henry V) who in the first half of the fifteenth century endowed Oxford with books to add to its meager collection and money for a room in which to house them all. The original Humphrey collection contained about six hundred manuscripts—said to be "the finest that could be bought"—but it was reduced drastically by the reform movement of the sixteenth century, and the bookcases were sold in the 1550s, there being no more immediate use for them. About fifty years later, Sir Thomas Bodley, a collector of medieval manuscripts, restored the library, and it reopened in expanded premises in 1602. The Duke Humphrey, which was fitted out with the stall system of bookcases, forms the core of a library that came to be flanked by two large rooms known as the Arts End and the Selden End. Together, these three great rooms form a capital letter H in plan, with the Humphrey, aligned in an east-west direction, being the cross member and the two ends the verticals of the H.

The Arts End was so called because its shelves held the folios associated with the Arts Faculty. (The other three faculties of the time were Theology, Law, and Medicine, each of which had its own collection of books.) The Arts End, which dates from 1612, was located in the eastern wing of the Divinity School, and it is remarkable in its contrast to the Duke Humphrey because it is fitted with what is thought to be the first use of the wall-system arrangement in England. Three large windows allowed light into the room, thus mak ing the wall system feasible; the largest was "the great east window," which aligned with the central aisle of the Duke Humphrey and let light in from the Schools Quadrangle. The two other windows were located in the narrower north and south walls of the room, and they also let in a good deal of light.

It is probably the case that the wall shelves were originally open, with the folios chained to them, but Sir Thomas did not want to chain octavo- or smaller-sized books, which were being produced and acquired in increasing numbers, and so these were kept under lock and key in the lobbies. The collection of these books grew quickly, however, for the library was a repository and as such was entitled to a free copy of every book printed in England.

To accommodate the smaller books, a gallery was added, and the pillars used to support it also then provided end posts for the benches that faced the bookshelves under the gallery, which was reached via stairs. Access to the gallery was restricted to the librarian and assistant to the librarian by means of a cage fitted with a door that could be locked. The gallery must have been a reluctant addition, for, as one can imagine, it cast shadows on the shelves and readers beneath it. These shadows are clearly present in an engraving of the Arts End

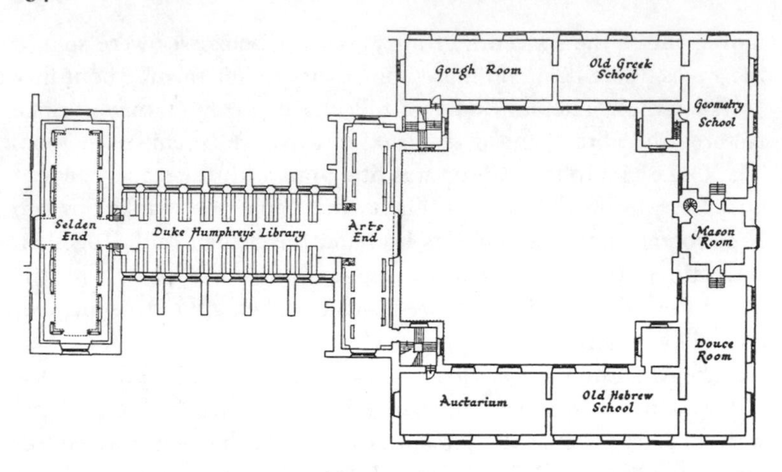

The Bodleian Library at Oxford is on the west side of the Schools Quadrangle. Duke Humphrey's Library formed the core of the library, before the collections of books in the Arts End and the Selden End were added. The library later expanded into rooms around the quadrangle.

dating from 1675 and in a watercolor of the room dating from 1843, in which there are clear shadows cast on the top shelves of books under the gallery.

These depictions of the Bodleian are also interesting in showing some library patrons in academic cap and gown, required dress for any undergraduate wishing to pass through the Schools Quadrangle, which was the only way to gain access to the library, a further measure of protection for the books. (To this day, student and graduate users of the library are asked to wear their academic dress, at least on their first visit.) The caps and gowns must have been alternately bane and blessing for these library users, for in the Duke Humphrey, at least, it was warm "in summertime, when honey-questing bees droned lazily in through ivy-framed windows opening on to Exeter College garden, but in winter it could be bitterly cold, and became dark as day and year wore on, for all artificial light was strictly forbidden under Sir Thomas Bodley's statute." The faculty and students may or may not have been comforted by the text *Dominus illuminatio mea*—"God is my light"—painted on each ceiling panel.

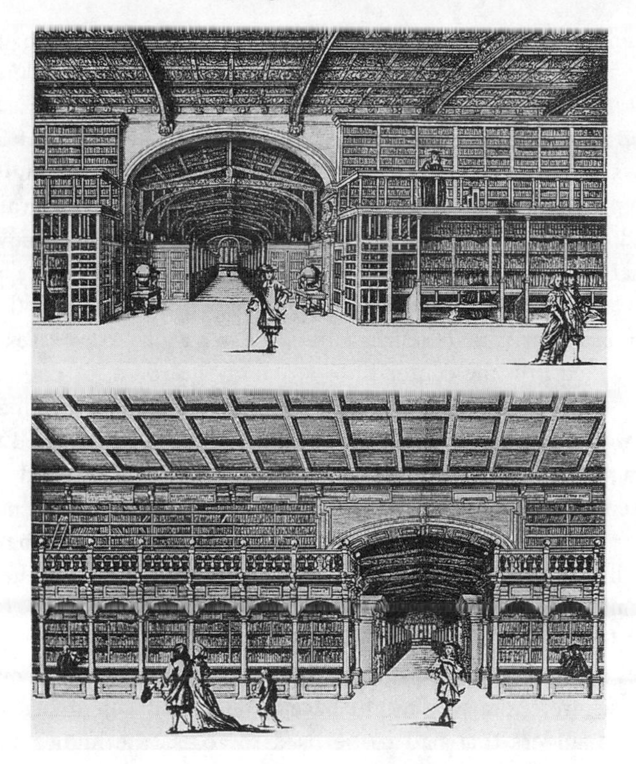

The Arts End (above) *and the Selden End* (below) *are shown here as they were depicted in 1675. Duke Humphrey's Library is seen in the background, with its stall system of shelves contrasting with the wall system of the ends. Note the students in academic dress.*

The Arts End was later joined through the Duke Humphrey by the Selden End, which was so called because its first wall shelves were built in 1634 to house the folios bequeathed to the university by John Selden, who was a lawyer, member of Parliament, scholar, and orientalist. His will ordered that the books be chained, but this presented a technological challenge for the wall system. In the stall system, where book presses were never much more than 8 or 10 feet long, even a single long rod could be managed when sliding the rings off its end to unchain a single book. If need be, the entire rod could be unlocked and removed; the wide center aisle provided plenty of room to accommodate its extraction, as it would have the rod's installation. When shelves were installed along a whole wall, however, a single long rod would have meant unmanageable inconvenience, to say the least, and so a series of shorter rods were used, as they often were even in the stall system, butted up against each other end to end on a number of rod supports and behind hasps located at convenient intervals and shaped to contain the rods. In time, shelves were added above the original ones and also beneath the desks to accommodate more books.

As long as a run of bookcases might be, without a gallery the wall system is a clearly inefficient use of space in a room. New cases could be projected into the center floor space of the room, of course, but neither their installation perpendicular or parallel to the original wall cases would be fully satisfactory. Perpendicular cases attached to the existing wall cases would cover some existing shelves and would involve major carpentry work in order to make the new cases harmonize with the old. Perpendicular cases a short distance from the existing wall cases would produce a hybrid situation that was neither wall nor stall. New shelving parallel to the existing wall shelves but an aisle's distance away would create problems with light on the original shelves and desks. Furthermore, depending on how tall the original wall cases were, the aisle might have to be pretty wide to accommodate any ladders that had to be used to access the higher shelves. Hence the gallery was the preferred way of achieving more shelf space.

Even with a gallery, the wall shelves in time become filled to capacity, and since they already go from floor to ceiling, they cannot be expanded, for there is only so much wall space and height in a room. Even when the drastic step is taken of converting barrel vaults to flat ceilings, as was done at the Mazarine, there is eventually nowhere else to go with the books than into the center of the room or into adjacent rooms. This, of course, presented new problems in the days before artificial lighting, but they were dealt with for the sake of having an extensive library.

As books continued to be acquired by the Bodleian at the rate of three or four thousand a year, shelf space was soon exhausted. That the shelves in the Arts and Selden Ends could not receive any more books as early as 1675 is clear from prints published in that year. The shelves are packed, with only a gap here and there where one has to assume that a book was removed—and expected to be replaced after

The rods used with a wall system of bookshelves had to have special hardware to lock them into place, since they could not be removed by being slid along the shelves, as they were in the stall system.

use. This situation of having nowhere else to put books is a common one among research libraries that continue to grow (as virtually all do). Even if there is new space constructed and fitted with new bookshelves, the problem of what books to put where is a perennial one.

Home libraries too present this dilemma to their owners, as discussed in the appendix to this book. In older systems of shelving, where books were located by press marks—that is, numbers assigned to books to designate their location in a particular partition or on a particular shelf—newly acquired books could just be put in the new presses. One common arrangement has been to shelve according to date of acquisition; another method has been to open up new shelf space throughout the whole collection by spreading out the books accordion-like every now and then, so that new titles cataloged according to subject matter can be added to the space so opened. This has presented special problems with the proliferation of books on computers and information science, which in the Dewey decimal system are classified at the beginning of the numbering system. Librarians have tended to underestimate how much space to open up for the next few years of acquisitions to this rapidly growing category of books.

Sir Thomas foresaw such problems of the "stowage for books" at Oxford, so he left money to construct a second story (the third floor) above the University Schools. The new space was used first as a picture gallery, however, and "it ranks as the oldest public gallery in England and formed the sole university art collection." Manuscripts did begin to be placed in the gallery as early as 1747, and some printed books were added later in the eighteenth century. The Picture Gallery began to be used for general book storage in 1824, but the rate of acquisition was such that "all new octavos other than serials were referenced into a fresh category called Year-Books." This meant, of course, a departure from the arrangement of books by subject, for "the octavo accessions of each year were simply placed in alphabetical order of authors' names." Quartos were shelved separately but similarly, suggesting that an optimal use of shelf space was being sought by having books of uniform size occupy shelves spaced just right for that size. (Interestingly, under this system the pressmark consisted of the last two numerals of the year acquired separated by a comma from a number indicating the book's place in an alphabetical ordering—24, 396, for example. Such a system ran the risk of creating a year 1900 problem long before the year 2000 problem that plagued digital computer systems in the late twentieth century.)

The Year-Books were shelved in book presses installed along the north wall of the Picture Gallery, "blocking up its north windows." A similar situation occurred in the southern wing of the Picture Gallery, where "its north windows, like those of the north range, had been blocked since 1831 by wall-cases." There were windows on the opposite, or south wall, left unobstructed to provide light, which of course would have been brighter than northern light, emphasizing that natural light was still an important consideration in illuminating bookshelves. But the decision to obscure the north windows in both wings of the gallery suggests very strongly that by the early nineteenth century the admission of the maximum amount of natural light had come to be seen as less of an imperative than the provision of shelf space.

It was not only in the Picture Gallery that light was sacrificed for shelving. In the late eighteenth and early nineteenth centuries, the Bodleian had taken over school rooms and other spaces on the first story, which was on the same level as the Duke Humphrey and its ends. The first addition to the library had in fact been the room known as the Auctarium, in which were fitted "tall wire-fronted bookcases, lettered A to Z." These bookcases "lined its walls and blocked its north and west windows," thus leaving only south-facing windows unobstructed to admit light and, it might be expected, fresh air and bees in the summertime. Cupboards were added below the bookcases in 1812. In 1828 the Astronomy School became annexed to the library, and shortly thereafter it was used to house the seventeen thousand volumes newly bequeathed to the university by Francis Douce. To accommodate and safeguard these books, shelves were installed in front of the room's west window, and an entrance into the room from the north was blocked, thus creating a cul-de-sac that could be entered and exited only through the Arts End.

In spite of the installations at the Bodleian, the wall system was not accepted all that readily in England and was referred to as "*à la moderne*" as late as 1703. Indeed, Sir Christopher Wren is believed to have been the first English architect to venture to design a library whose windows "rise high, and give place for the deskes against the walls." This was in Cambridge's Trinity College Library, completed in 1695, some three decades after Wren had visited France and admired "the masculine furniture of the Palais Mazarin." Nevertheless, in the space that he designed (191 feet long by 40 feet wide), Wren did not go wholly to the wall system, for he declared,

> The disposition of the shelves both along the walls and breaking out from the walls must needes prove very convenient and gracefull, and the best way for the students will be to have a litle square table in each celle with 2 chaires. The necessity of bringing windowes and dores to answer to the old building leaves two squarer places at the endes, and 4 lesser celles not to study in, but to be shut up with some neat lattice dores for archives.

The combination wall and stall system that Wren adopted is a familiar one in the reading rooms of many smaller libraries today. However, it did not solve the problem of libraries growing into spaces not originally intended for them. When I first came to Duke University in 1980, the Engineering Library was housed in "Old Red," as the red brick main Engineering Building is affectionately known. When this building was constructed in the late 1940s, by design the library occupied a central location on the second floor, above the auditorium. In the grand tradition of libraries, there were many windows around the room—which, against tradition, was squarish—and between each pair of windows were nicely finished oak bookcases, with matching

The Wren Library of Trinity College, Cambridge, was designed by Christopher Wren. Its shelving can be seen to be a combination of the stall and wall systems.

wainscoting carried into the window areas. The room must have served as a wonderful study hall for the engineering students of the time. But as the book collection grew, so did the need for space.

The library appeared originally to have had a librarian's desk out in the open behind the counter that was situated in the center of the room, directly in front of the entrance. The librarian's desk was in time enclosed in glass, so that the librarian worked in a fishbowl of sorts, but with a degree of aural privacy. The librarian whom I came most to know kept the book and journal collection in impeccable order, almost to the neglect of—but not the abandonment of—his own space. After his desk became full, the floor accumulated piles of catalogs, pending orders, and assorted pieces of paper that served patrons in one form or other. In time, the piles began to tumble down upon one another and coalesce, so that the glassed-in office resembled an uncleaned fishtank in which the detritus was flowing with the movement behind the glass and was accumulating in especially high piles in the tank's corners. (The construction of a new library in a new building enabled the librarian to move into a new office and to begin life anew with a visible floor, but it was not long before it too became covered with papers of all kinds.)

For all the disorder in his office, however, the librarian could always find what he needed, including space for books that were added to the collection. Rooms on either side of the library had already been annexed by his predecessors, and later even more remote but still contiguous rooms, so that to find a book one had to wend one's way through rooms as if in a railroad flat. As in the Bodleian, the only way into and out of these rooms was past the main desk, and so the security of the library's collection was ensured. When the new engineering library building was built, the old library rooms were reclaimed to be the classrooms and laboratories that were so needed. There is no great improvement in convenience in the present library, however, for the books are still more or less in the same order on the bookshelves. The shelves are newer and more uniform, and there are wider aisles and generally better lighting. But once one has made one's way to the vicinity of the section of shelves that contains the object of a visit, the shelves properly fade into invisibility, as good infrastructure is supposed to do, and all one sees are the books. And books often evoke memories.

My first bibliothecal recollection is of a storefront branch of a New York public library. What the store was before it became a branch library I never knew, but it must have been a business that required a lot of floor space. Because the room was so large in area, it also had a large perimeter, three sides of which were covered with bookcases. Since it was a storefront in the middle of a block on a commercial street, there were no side windows, and the only source of natural light was through the large plate-glass windows that once might have been decorated with SALE or GOING OUT OF BUSINESS signs but that became filled with colorful dust jackets and signs of the season cut out of construction paper. The wide-open space in the middle of the room was filled with tables and chairs, at which children studied after school. The librarian's desk and counter were located directly in front of the entrance, and so in some ways it was little different from the Bodleian's Arts End, Duke's old engineering library, or countless other libraries large and small.

Paradoxically, as I recall, there were always new books appearing on the library shelves, but the shelves never seemed to get fuller; for as long as I recall patronizing it, my childhood library encountered no need to expand its bookshelf space. Whether this was because increasing numbers of books were being checked out or because the library regularly discarded old books, I did not know or even think to wonder at the time. However, in retrospect, the most likely explanation for the seeming exception to the rule that libraries always need new shelf space is that the main library could keep a fresh supply of books in its branches by rotating their collections. In any event, had there been a need for more shelf space, the storefront certainly had it, and shelves could have been placed perpendicular to the wall reaching into the great open room like jetties into the ocean. Alternatively, bookcases parallel to the existing side-wall cases could have been installed, without blocking the light from the street but contracting the space ever so slightly, as in a horror or adventure movie in which the walls begin to move inward upon the trapped hero and heroine. Although there was some space above the library's wall cases to add a gallery, this would have been pretentious in this modest storefront.

A gallery did not seem pretentious in Professor Henry Higgins's expansive study in *My Fair Lady,* however; nor did it seem so in many a Victorian mansion, allowing for the fact that the mansion itself might not have been considered inordinately pretentious in its context. Although galleries have not been so common in late-twentieth-century studies, floor-to-ceiling bookcases have been, and when the ceilings are extra high, as they are in the lofts that have become so fashionable, there is need for some means to reach the books on the uppermost shelves. The ladder has been the means of choice.

As the gallery appeared first in institutional libraries, so apparently did the ladder. In the late nineteenth century, Melvil Dewey reported having seen in England tall ladders enabling books to be retrieved from shelves as high as 25 feet off the floor. He also said that prior to moving into a new building in which no shelf was more than 7 feet off the floor, a ladder weighing 75 pounds was the "sole means of access" to library shelves 24 feet high. He also described a form of ladder that he saw first in Birmingham and later in the Locust Street branch of the Philadelphia Library, where it was installed in a new building in 1880:

> A common, light but strong ladder has at the top, bronze metal lether-padded hooks, as if each side ended in a half oval metal claw, 3 to 4 cm wide. Opposit the edge of the top shelf runs a gas pipe (¾ inch outside) about 1.5 cm from the front of a shelf. This is firmly fastend to each upright by a small galvanized iron bracket. The pipe is only far enuf from the front edge of the upper shelf to admit redily the lether-cuverd metal hooks. It appears from the floor, therefore, as a round edge to the shelf, projecting about 4 cm. Being so near the shelf and exactly opposit its edge, it is therefore not in the way of putting books on the shelves, and yet is a firm and safe support for the ladder, which hangs on the pipe, and is slid along wherever wanted. The hook, being a full half oval, cannot slip off with the weight of the ladder holding it on, and one may lean to the side as far as he can reach without danger of the ladder falling.

Such a ladder, being hung on the gas pipe, could be used in a nearly vertical position, thus not extending out very far from the base of the bookcase. In the case of the Philadelphia Library, the ladder was almost 15 feet long but its foot was only 2 feet 8 inches from the shelves. As originally installed, the ladder's bronze hooks on the iron pipe produced an annoying metal-on-metal sound whenever the ladder was used, "but this was remedied later by covering the metal hooks with lether." Ever the stickler for detail, Dewey recognized that when the ladder was not in use it projected out into the aisle, and so he recommended that a short rod be installed one shelf higher than the gas pipe. When suspended from this rod, the ladder "hangs flat against the case wholly out of the way of passers, and if books ar wanted they are redily reacht between the rounds," or rungs, of the ladder.

In time, bookcase ladders came to be fitted with wheels, thus eliminating the need to cover their hooks with leather, which did wear out. At the top, the wheels rode on a pipe-like rail, whose brackets were designed so that they did not interfere with the sideways progress of the ladder. The bottom of the ladder was also fitted with wheels so that it did not have to be lifted up or scraped along the floor to be moved from section to section of the bookcase. (Not all book ladders have or need wheels, however, nor did the ladder wait for Dewey to discover it. In the Wren Library today, for example, the top of the ladder is fitted with a wooden frame that leans against two shelf edges for support and allows a book to be removed through the frame.) Where high bookcases have been installed in converted lofts, the exposed water and gas pipes already in place have sometimes provided handy places on which to rest a ladder. The artist Michele Oka Doner uses a ladder to reach the upper shelves of the striking black bookcases she has installed in a form of wall/stall arrangement in her New York City loft.

In the late nineteenth century, if a gallery or ladder was not available, books in high wall cases could be retrieved with Congreave's Book Reacher, which reminded Dewey of an American apple picker. But for those who grew up in the mid-twentieth century, just as the rolling library ladder might remind us of the ladders once used in shoe stores when the stock was kept on high shelves surrounding the fitting room, the Congreave reacher might be likened more to those grasping contraptions that grocery-store clerks used to get boxes of cereal down from high shelves before the self-service supermarket, with its uniformly low but not always reachable shelves, became so popular. Whatever it is likened to, Dewey described the book reacher as follows:

> On a pole is a pair of metal jaws coverd with rubber and workt by a rod or chain and lever. These jaws ar placed at the back of the book, and by pressure at the lower end the book is tipt out of its place into the jaws, so that it can be lifted down safely. It is put back in the same way. To aid in reading the titles of the higher books a magnifying glass is attacht. One would naturally be sumwhat skeptical as to the practical value of such a machine, but those who hav used it giv their testimony as to its wurth. . . .

Most home libraries are located in rooms that do not have ceilings high enough to present problems in reaching the uppermost shelves of bookcases. That is not to say that the desire is any less acute to have a room lined with books from floor to ceiling, sometimes even when there are not enough books in the household to press a flower. The phenomenon of trompe l'oeil wall painting depicting books on shelves appears to have been the subject of an especially playful fad in the mid-nineteenth century. The duke of Devonshire, "desiring to construct a door of sham books, for the entrance of a library staircase at Chatsworth," sought help in inscribing the painted spines. The duke had grown tired of such titles as *Essays on Wood,* and he engaged the humorist Tom Hood to help brighten up the door. Hood's contributions included Lamb's *Reflections on Suet,* John Knox on *Death's Door, On Sore Throat and the Migration of the Swallow, Cursory Remarks on Swearing,* and *The Scottish Boccaccio* by D. Cameron. A contemporary's library door was decorated, near its hinges, with *Squeak on Openings* and *Bang on Shutting.* Books-on-bookshelves wallpaper has replaced door painting for the most part, but not with such a sense of wit. In the late twentieth century the image of books on shelves began to be worn by men when it became commonplace for them to sport neckties with strong horizontal patterns and large bold depictions of everything from gorillas to school buses. Such ties seemed to be especially attractive to wallflowers.

In spite of its need for accouterments to reach and retrieve books, the wall system would become the standard arrangement of shelves in the reference and reading rooms of large institutional libraries, and it would predominate even in private home collections. At first, the shops of booksellers would also favor the wall system of shelves, but in time, as their stock grew with the growth of literacy and the general growth of commerce, the open space in the middle of rooms began to be coveted for practical and commercial reasons, and sometimes even for coffee shops and cafés.

CHAPTER EIGHT

Books and Bookshops

Setting moveable type—letter by letter, word by word, line by line, page by page—was certainly little different than copying out a manuscript, but once that type was set, its reverse image could be inked and pressed time after time after time onto blank sheets of paper and transform them in one fell swoop into printed pages that could be gathered into books. The essential technology to do this was in place by the middle of the fifteenth century, thanks to Johannes Gutenberg's innovative method of casting metal type and his development of an ink that would adhere to it and to paper, which enabled Gutenberg to typeset, print, and publish his 42-line Bible in Mainz, Germany, in the early-to-mid-1450s. All books that were produced by this new technology up to the year 1501 are known as *incunabula,* which is Latin for "things in the cradle," and an *incunabulum* is an individual book that came out of the infancy of printing. The Latin was Englished in the mid-nineteenth century to "incunable," with the straightforward plural "incunables," a word that replaced the older English term "fifteeners" for books printed in the fifteenth century.

Incunabula, being books of a transitional period, often owed much of their appearance to manuscripts, including multiple columns of text per page and initial letters added by hand or printed in a contrasting color of ink. Estimates vary, but the total number of incunabula that survived to the nineteenth century has been thought to be between fifteen thousand and twenty thousand. The number of each title printed varied, as it does today, according to expected sales, but several hundred copies often constituted an edition.

Unlike in the Middle Ages, when "a great book might be available in a hundred manuscript copies, and read at most by a thousand people," after the middle of the fifteenth century a book "could be available in thousands of copies and read by hundreds of thousands of people." It has been estimated that in the sixteenth century in Europe alone there were more than one hundred thousand different books printed. If it is conservatively assumed that there were on average as few as one hundred copies of each book (print runs of several hundred were not uncommon in the fifteenth century), ten million individual copies of books were available to Europeans. (Some estimates are ten times this.) Thus, by one very conservative estimate, "the power of the printed word increased a hundredfold the power of the written word." Furthermore, more books meant more readers, which translated into more writers, which in turn led to the production of still more books. And more and more books meant an increasing need to find more and different ways to store and display them, including in shops where they were sold.

The shops of a printer and a bookseller were pictured in a 1499 edition of *Danse Macabre*, an early illustrated printed book published in Lyons, France. In the depiction of the bookseller's shop, the books on the shelves are all horizontal, and none has its spine facing outward. Another early picture of a bookseller's shop has been identified by Graham Pollard, who has written definitively about the changes in the style of bookbinding that occurred between the sixteenth and nineteenth centuries. The illustration is in the book *Orbis Sensualium Pictus,* which was "the first picture-book ever made for children and was for a century the most popular text-book in Europe." Its author was the Czech theologian and educator Jan Amos Komensky, who wrote under the name Johann Amos Comenius. His text was published in London in 1655, just about the time that Samuel Pepys might have begun frequenting the bookshops in that city. According to one of Pepys's contemporaries, speaking of a bookseller, "He keeps his stock in excellent order, and will find any book as ready as I can find a word in the dictionary."

How such order might be kept is suggested in the engraving in Comenius, which shows a shop interior, lined with shelves and having a counter on which a lectern holds a book being read by a person, perhaps a patron like Pepys. The shelves have two distinct arrangements. About two-thirds of those shown are fitted with what have been described as "filing cabinets or bins," which appear to be labeled. It

The shops of a printer and a bookseller were depicted in a 1499 edition of Danse Macabre, *published in Lyons, France. The books are clearly shelved horizontally, and none has its spine facing outward.*

has been estimated that the largest of the many drawers shown in the illustration, those near the floor, measure about 18 inches wide and 2 feet tall, and the upper drawers are of diminished size. Since they are closed, it is not possible to tell with certainty what is in them, but it is unlikely that they held books in the shape and form we know today.

The booksellers of the later seventeenth century were not likely to carry bound books at all, for it was then the custom to buy one's books in loose quires, or gatherings of printed sheets. These could be folded into sections—often called signatures because of the letter or letters that were printed at the bottom of the first page of each section so that they would be assembled in the proper order to make a proper book—to be bound in whatever material the book buyer chose, but often by a different person than the one who sold the printed material. Depending on how many times the original sheets were folded, the sections when bound would form a folio, quarto, or octavo, designating one, two, or three folds, respectively, giving two, four, or eight

A bookseller's shop is depicted in the first illustrated textbook, Orbis Sensualium Pictus, *published in 1655 by the Czech theologian and educator Comenius. The labeled drawers lining the walls to the left are believed to have contained books in the form of unbound printed sheets. Bound books on the right are shelved fore-edge out.*

leaves per gathering. Since each leaf consists of two pages, front and back, folios, quartos, and octavos have, respectively, four, eight, and sixteen pages per gathering or signature. A duodecimo, abbreviated "12mo," which is pronounced "twelve-mo," has twelve leaves per gathering, giving a signature of twenty-four pages. Still smaller books came in 16mo and 32mo sizes. No matter what the format, the exact dimensions of the finished book depend on the size of paper with which the printer started. The book's thickness depends on how many signatures the book contains, which of course reflects the number of words in the text and the size of type in which it was set.

Just as we cannot be sure of the contents of the closed drawers in an engraving, so neither can we be sure of their full inside dimensions. However, in the case of Comenius's illustration, it is easy to imagine that the larger lower bins held folded folio sheets, and that the smaller upper bins held sheets for quartos, octavos, and smaller-format books. The labels on the bins may very well have come from the printed sheets themselves, for it is not uncommon to find books printed in the latter part of the seventeenth century "where the printer has printed the title of the book vertically on a leaf that would otherwise have been blank." It has been speculated that these titles were to serve as labels that could be cut out and pasted on the spine of a plain calf binding or "inside the cover to form a flap over the fore-edge" if the book was placed spine inward on the shelf.

The idea of printing the title, often the abbreviated title, of a book on an otherwise blank sheet survives in the form of the half-title, also known as a fly-title or bastard title. This is often the first page one sees in opening a book, and it appears to have developed from the frequent practice of leaving the first leaf blank to protect the title page from dirt and damage before binding. Printing on this leaf something to identify the unbound sheets appears to date from the latter part of the seventeenth century. Sometimes these half-title pages were removed before binding, but sometimes they were not. Confusion as to what to do with the half-title page persisted into the mid-nineteenth century. According to a "hint to book lovers" from that period, "Never allow the binder (as is often done) to remove the bastard (or half) title; it is part of the book."

The cabinet in the foreground of Comenius's illustration has large drawers fitted into it, and it is easy to imagine that these might have held the sheets of the largest folios. Behind the cabinet, at which the reader is standing, are shelves of varying heights that appear to hold bound books of varying sizes in vertical positions, with their fore-edges out. Comenius's primer went through several subsequent editions, with the illustration of the bookshop remaining unchanged as late as 1705. However, in the 1777 edition, the shop interior is shown from a wider perspective, and the arrangement of stock on the shelves is much different. Here the shelves appear to be uniformly filled with bound books, arranged clearly with their spines out, as had by then come widely to be the practice.

Long after the shelving of books became regularized, the lack of uniformity in size among them was a persistent problem. My book on the pencil was published in a format suggestive of its subject, a format that was taller and narrower than the usual octavo. This worked fine to distinguish the book and draw attention to it when it first appeared in hardback, but the unusual size proved to be a problem when the book

In the 1777 edition of Comenius's book the illustration of the bookseller's shop was updated to show that the shelves were stocked with bound books with their spines facing out.

was to come out in paperback, a format in which there is much more uniformity of size to conform to display racks. In the end, the intended paperback publisher did not want to handle the book, because it would have had to be typeset all over again to bring its pages into conformity with those of a standard-size book, and so the publisher of the hardback brought out the paperback in the original, unusual format.

Problems of book size have been felt especially acutely among librarians, some of whom would go to elaborate lengths to address the matter. The New York Public Library, which dates from 1895, made "a careful study" of how to classify books. Octavos were defined as those up to 11½ inches tall, quartos were between 11½ and 19 inches tall, and folios were those in excess of 19 inches tall. If the standard height of a section of bookshelves was taken as 7½ feet, it could be fitted with no more than seven shelves and still allow a tall octavo "to fit snugly." Another shelf might be squeezed in when shelving fiction, but not nonfiction, for too many volumes of the latter would have to be "turned down" on their fore-edges.

Melvil Dewey worried about the size of bookshelves, as he seems to have worried about everything in libraries, and he believed that "the common error is waste of space by giving too great depth to shelving." He argued that 80 percent of the books in a circulating library were octavo size, which he abbreviated as O. In a passage demonstrating once again his zeal for spelling reform, he wrote:

> The common O is only 15 cm (6 inches) wide. Large O ar seldom over 17.5 cm (7 inches), so that a shelf 20 cm (8 inches) deep allows liberal margin for books and for a little air space. It is common to make shelves 10, 12, and even 14 inches thruout the library. We hav seen them as deep as 20 inches, wasting both lumber and space, and annoying the shelf clerk constantly by the loss of books, which get pusht back into the vacancy behind the row in front.

Concerns over the different sizes of books were beginning to arise even back in the seventeenth century. Bookshelves were not then ablaze, however, with colorful bindings, garish paperback covers, or creative dust jackets. Sometimes the quires of different works were bound together, perhaps to save money, or perhaps to gain a more uniform thickness of book for the library to which it was to be added. As late as the nineteenth century, however, book collectors were advised never to bind "a quarto with a duodecimo—the latter is sure to fall out." Even though such rules were generally followed, volumes made up of what may seem to be totally unrelated titles can still be found when older books are consulted in a large research library. When we unknowingly request such a volume in the library, the book that is finally presented to us as the title for which we asked may appear at first to be something else entirely. The title on its spine may not be the one we expected, and opening up the book to the (first) title page may do little but confirm that we have been handed the wrong book. (The nineteenth century saw an urge to "disbind" such multiple works and rebind the most desirable individual parts in handsome leather whose "backs were so narrow that the title had to be tooled in letters too small to read.")

Just as codices typically were not bound by the scribe, so, as has been noted, early printed books were not generally bound by the publisher. Though printers did bind and sell their own books, book production and distribution soon became distinct, and a printer could not legally sell books directly to the public unless registered as a stationer. During the sixteenth century, stationer-booksellers typically maintained workshops in which binding was carried out. In the next century these merchants continued to act as middlemen, but the actual binding came to be done by master binders. Those books that were bound directly by the stationer or by the binders he employed became known as trade bindings, and were more or less common, as are the bindings in which most books are issued today. There were alternatives, of course, and "rich, private collectors continued to have their books bound in a more sumptuous manner, using as a rule damask and velvet rather than leather." Such was far from the standard stock in bookshops, however.

The booksellers of Pepys's time were often also publishers who carried their own titles, and even when they were not, each of them tended to have a unique stock. Thus Pepys frequented a number of shops around London, and he took what purchases he made to a separate binder for making into what we would consider a finished book. How the book was bound depended upon one's budget and one's taste, which was of course subject to change, and seventeenth-century buyers tended to have their own bookbinders the way we have today our own plumbers, doctors, and stockbrokers. In returning to the same bookbinder time after time, book collectors like Pepys could have their volumes bound uniformly, something we tend to see in old libraries, as well as in modern ones that feign antiquity.

If one wished perfectly matched bindings, as was often the case, one could have all one's books bound by the same binder at the same time. As late as 1665, Pepys got his books bound through a stationer or bookseller, each of whom is believed to have acted as a middleman between the book owner and bookbinder. The diarist wrote in January of that year: "Up, and by and by to my bookseller's and there did give thorough direction for the new binding of a great many of my old books, to make my whole study of the same binding, within very few." The binding was apparently finished within about two weeks, for then Pepys was writing, "Down to my chamber, among my new books, which is now a pleasant sight to me, to see my whole study almost of one binding." A year and a half later, however, Pepys was dealing directly with his own binder, for in August 1666 he recorded that he had gone "to Paul's churchyard to treat with a bookbinder to come and gild the backs of all my books to make them handsome, to stand up in my new presses when they come." Before long the presses were installed and admired, but soon they were filled. Within only a few weeks Pepys was again worrying about space in his bookcases:

> All the morning setting my books in order in my presses, for the following year, their number being much increased since the last, so as I am fain to lay by several books to make room for better, being resolved to keep no more than just my presses will contain.

In time, Pepys did get more presses, of course, and a visitor to his library in 1702 found it "arranged in 9 cases finely gilded and sash glassed." Furthermore, according to the visitor, the books were "so well ordered that his Footman after looking at the catalogue could lay his finger on any of 'em blindfold." Indeed, each case and shelf was numbered, with the front of each pair of double shelves being designated "a," and each rear one "b." To find a book, one determined its unique number from the catalogue; a "table" of positions then led one to the proper case, shelf, and location along the shelf holding the book. Pepys's books were shelved spine out, as had come to be the fashionable thing to do, and many of the bindings were tooled and gilded. Volumes at the beginning and end of a shelf tended to have their catalogue number discreetly affixed near the top of the spine, to aid in the arrangement and finding of books.

But no matter how handsome a finely bound book could be, not everyone wished to spend as much money as Pepys did on a binding or on the press in which to display it. Even Pepys himself distinguished between books to be bound for posterity and others:

> To the Strand, to my bookseller's, and there bought an idle, rogueish French book, which I have bought in plain binding, avoiding the binding of it better bound, because I resolve, as soon as I have read it, to burn it, that it may not stand in the list of books, nor among them, to disgrace them if it should be found.

William Dugdale, the English antiquary who has also come to be referred to as "the grand plagiary," had a significant reputation among mid-seventeenth-century scholars with whom he interacted. Though much of Dugdale's work depended upon the scholarship of others, "he had a special skill in bringing to the point of publication work which other men had left in the form of disordered notes, and there is no doubt that he here performed in many cases a signal service to scholarship." Among Dugdale's most recognized works is one he published with Roger Dodsworth, who had assiduously searched records that led to the material in the *Monasticon Anglicanum,* which presents many original documents relating to English monasteries.

It is a copy of the *Monasticon* that occupies a prominent place on the table beside which Dugdale is sitting in the portrait engraved by Wenzel Hollar in 1656. The book is identified by its clearly readable title on the front cover, which is one way in which bound books were marked at the time. On the table is also another book, with its fore-edge facing out and lettered to clearly identify it as *The Antiquities of Warwickshire Illustrated,* a book that was wholly Dugdale's and published the year after the first volume of the *Monasticon* appeared.

What draws our eye here, however, is the stark modern-looking bookcase in the background, over Dugdale's right shoulder. On its shelves is a jumble of rolls and books, the latter including both bound and unbound ones, all appearing to have been thrown onto the shelf with little regard for their orientation or care. None of the bound volumes is facing spine out, but what is perhaps most interesting in the present context is the presence and condition of the unbound books. These show how books might have been bought from the bookseller at the time Pepys was purchasing his own books and show how different readers treated what they bought. The pages in Dugdale's case appear to be curled up and folded over, the way many of us might read a disposable magazine today, and perhaps the way Pepys read his "rogueish French book." There is clearly no care being taken by Dugdale with these works, whatever they are, and it would seem unlikely that he had any intention of ever binding them. They may be books that were useful to him, even if only as reading for diversion, while he was working on whatever new project required the pen and ink before him on the table. For all we know, when that project was done, he might clear out his shelves to make room for a whole new assortment of gathered but unbound books.

That all readers do not keep or even care for every book they own was brought home to me once by a professor who claimed that he tore out the pages of paperbacks as he progressed through them. The rationale was to eliminate bookmarks, which might become dislodged

The condition of the books on the bookshelves in the background of a 1658 engraving of the English antiquary William Dugdale suggests that he did not have all his volumes bound. One of the bound books on the table has its title inscribed on the fore-edge.

anyway, and, as I recall, to lighten the burden of what baggage he carried (the latter explanation being fraught with metaphorical meaning). It was all done, I believe, to shock young graduate students who the professor thought took themselves altogether too seriously. However, it converted few if any of us to adopt the practice, even if it was said to have been done by the nineteenth-century English chemist Sir Humphry Davy. If I am recalling the story correctly, Davy excused the practice of disfiguring his books because he did not believe he would have time in his life to read anything twice and so would not allow himself to be tempted to do so. According to the daughter of Clifton Fadiman, the longtime judge for the Book-of-the-Month

Club, "in order to reduce the weight of the paperbacks he read on airplanes," her father "tore off the chapters he had completed and threw them in the trash." Napoleon Bonaparte, whose coach contained a bookshelf, reportedly tossed books out the window when he was through with them.

The practice of caring for books more as things to be read than as objects to be displayed is also recorded in Joshua Reynolds's famous 1775 portrait of Samuel Johnson. There are no bookshelves behind Dr. Johnson in *Blinking Sam,* but he is shown squinting at an unbound book that he is holding folded over like a magazine or newspaper in his hands. Such a "book" would hardly be in fit condition for binding after such handling, but evidently some omnivorous readers of the eighteenth century were habitually so eager to devour their latest purchase that they were unwilling to wait the several days it would take to have it bound. Unbound sheets, as the unfinished book is known to the trade, continued to be published by Oxford University Press into the latter part of the nineteenth century, but the practice died out as binding became increasingly mechanized and expected by booksellers and book buyers to be done by the publisher.

There is a striking contrast between the unbound sheets and the bound books in the portrait of Dugdale. The former are swollen out at the edges, while the latter lie flat and close square. In this regard, the engraving may not be true to reality, for some old bound books tended to swell as their pages absorbed moisture. The part of the book near the spine was, of course, held together by the stitching and binding, but the unconstrained fore-edges of the book would fan out, especially if they had been stored in a damp environment or suffered water damage. As has been noted, one of the main reasons that early books of value were fitted with heavy boards and also clasps and other fastenings was to keep their pages flat, for parchment and vellum tended to wrinkle when not kept pressed together. The heavy, thick wooden boards alone added weight to the book when in a horizontal position, thus enabling gravity to work as if the volume were in a binder's press. This was also one of the reasons that unclasped books were kept flat or on gently sloping shelves. Even when wooden boards provided a pressing force, fastening together the front and back of a closed book kept a volume neat and its pages flat, and enabled the book to be stored vertically as well. (Since paper is less sensitive to changes in humidity than parchment, books printed on paper "did not require the weight of wooden boards to keep them flat," and pasteboards, made by pasting together multiple sheets of paper to develop stiffness, came to replace wood for book covers. Problems with humidity remain with some paperback books, however, and the coated cover stock curls up when the air gets damp the way the bimetallic strip in a thermostat does when it experiences a temperature change.)

In an afternoon of wandering unattended about the rare-book stacks of one university library, I encountered numerous examples of sixteenth- and seventeenth-century books whose fore-edges were two or three times thicker than the spine. Since less elaborately bound books of that era were not necessarily shelved snugly between other books, as the portrait of Dugdale demonstrates, they were allowed to absorb moisture and could not easily regain the flat profile that more carefully cared-for books have retained. In addition, many once-clasped books have lost their hasps due to wear and tear, thus allowing their fore-edges to swell. Where a book's clasps still function—and there are plenty of examples of these also extant—the body of the book, though five centuries old, still retains its shape.

Decisions whether or not to bind sheets, as well as earlier decisions whether or not to provide a bound book with fastening devices, were also matters of economics. Gabriel Naudé, who was librarian to Cardinal Mazarin and who collected the forty thousand volumes that constituted the Bibliothèque Mazarine, wrote of the economy of bookbinding from his French perspective in 1627. In John Evelyn's translation:

> As to the binding of books, there is no need of extraordinary expense. It were better to reserve that money for purchasing books on the largest paper or of the best edition that can be found; unless that to delight the eyes of Spectators, you will cause all the backs of such as shall be bound in sheepskin, as well as in calfskin or morocco, to be gilded with filets and some little flowers, with the name of the Authors.

Naudé here touches on what has been identified as "the three passions for identification, for display, and for uniformity," that led many subsequent owners of books to add "many embellishments to the backs of books already bound."

A change in bookbinding practice appears to have occurred around 1700, at least in England, when the customary manner of issuing an author's works in a single large volume was replaced by the new fashion of publishing them in multiple volumes. For example, in 1692 the plays of Ben Jonson were brought out in a single folio edition. In 1709, on the other hand, an edition of Shakespeare's works comprised nine octavo volumes. As binding became somewhat customary among booksellers, with agreed-upon prices published each year, the shops offered books priced to include a standard binding, which usually meant plain sheep or calf, without any lettering or decoration on the spine or elsewhere. As long as a shop's stock was small and works appeared in single volumes, there was little problem in keeping a variety of unmarked books organized. However, with the proliferation of multivolume sets, the possibility of a book buyer walking away with two copies of volume IV and none of volume V was very real. Given that the purchaser might not discover the error until a good deal of reading in the set had been done, the problem with the balance of the stock might not be discovered until well after the bookseller had had the opportunity to sell the last remaining set (and was confronted by a more careful book purchaser with the fact that there was no volume IV but two copies of volume V on the shelf). It was this kind of occurrence that might have led booksellers to imprint volume numbers—and often nothing else—on the spines of commonly bound books.

As the number of books stocked by a shop increased throughout the eighteenth century, the problem of distinguishing volumes, whether of sets or individual titles, became more important. Because of the practice of rebinding books that were originally bound in common calf or other less desirable binding for a subsequent owner's needs or desires, there is little reliable or conclusive evidence of exactly how and when it became standard practice to imprint titles, authors, and other identifying information on a book's spine. However, when the imprinting of volume numbers on the spines of multivolume sets was instituted in the first half of the eighteenth century, the inclusion of author and title on spines soon followed as a popular practice, with the author and title often accompanied by the year of publication. This was especially true of books that were bound in a style beyond the stock bindings.

As illustrated in Denis Diderot's *Encyclopédie* and as Adam Smith's *Inquiry into the Nature and Causes of the Wealth of Nations* made clear, the greater part of the eighteenth century, at least, was a time of much hand work and division of labor. Virtually everything, from pins to pencils to bound books, was made a step at a time by a workforce aided little by any power other than human muscle. The steam engine was developing rapidly during this time, of course, but at first it was used primarily for pumping water out of mines and not for driving manufacturing machinery. In time, steam did drive boats and, later, railroad engines, and increasingly served as a source of power for machinery of all kinds. The book industry was not to be left behind in the general fascination of the nineteenth-century inventors to mechanize and power just about everything that moved.

Although cloth binding as we know it was first adapted to bookbinding in 1823, "a style of binding uniform for all copies of the same book" did not appear until around 1830, when machinery was introduced to letter the clothbound cases that could be fitted over the printed guts of a book. This development ushered in a new chapter in the way books were made and sold. Whereas the bookseller would bind or have bound, by hand of course, only as many copies as were likely to be sold in the immediate future—a form of just-in-time manufacturing—with the advent of machinery the publisher itself began to bind an entire edition of a book in the common style of the time. Bookshops no longer needed to stock loose sheets, and so their shelving requirements changed to resemble more those of a library, where books had for some time been shelved vertically, with their lettered spines out. Where the private library had matched multivolume sets, if not matched bindings for all of its books, the bookshop had runs of multiple copies of the same book. The need to build shelves to display the books grew with the number of titles published and stocked.

Bookshops at the end of the second millennium come in many shapes and sizes, of course, and among the most talked about are the large superstores. These emporiums appear at first to be as far from the shops of the seventeenth century as the laser printer is from the printing press that produced the Gutenberg Bible. Although known as chain stores, these shops have books that are only metaphorically chained to the store through the magnetic tags that trigger alarms if carried through the shoplifter detectors at the store exit before being desensitized at the checkout counter. But books may be freely carried around the store to be read in an easy chair or over a cup of coffee that can be bought on the premises. In this regard, modern superstores do recall the bookshops that Samuel Pepys frequented, where "seats were provided so that customers could sit and read for as long as they liked":

> Pepys found these shops useful as places in which to wait between appointments or to meet his wife or friends, in the meanwhile occupying the time by reading or chatting with the booksellers or other customers. Even the more eminent booksellers seem to have been generally present in their shops and, being educated and knowledgeable men, their conversation was the easiest way of learning what was going on in the literary world, in the absence of regular book and theatre reviews as we know them today.

One thing that tends not to distinguish modern superstores from the smaller ones known as independents is the shelves on which the merchandise is displayed. Bookstore shelves continued to evolve in the late twentieth century just as they had throughout their history. Among the more popular shelf units appear to be the freestanding ones that are seldom higher than eye level, and are often well below it. These are convenient when browsing with a friend or spouse, so that each can go his or her own way without losing sight of the partner when either one wants to share a newfound book, or to leave. In bookstores with taller shelves, such as are found in my local Barnes & Noble superstore, I can spend an inordinate amount of time looking for my wife to show her a book I have found or to see if she is ready to leave the store. If I happen to look down an aisle just as she is rounding the corner at the other end, I can completely miss her, as I can if I look left when she is walking right down the opposite aisle. Merchants can also better watch over their customers when shelves do not rise over shoulder height, and the arrangement also gives a sense of openness to the store.

Many a bookstore shelf is not quite horizontal but tilted slightly up toward its front so that the books can be arranged cover out without tending to fall into the aisle. This slight slanting of the shelves—vaguely suggestive of the sloped lecterns used by medieval and Renaissance scholars to hold and display the decorated front covers of their books—also enables customers to see better the books on lower shelves, because the covers or spines are tilted up a bit. Even this

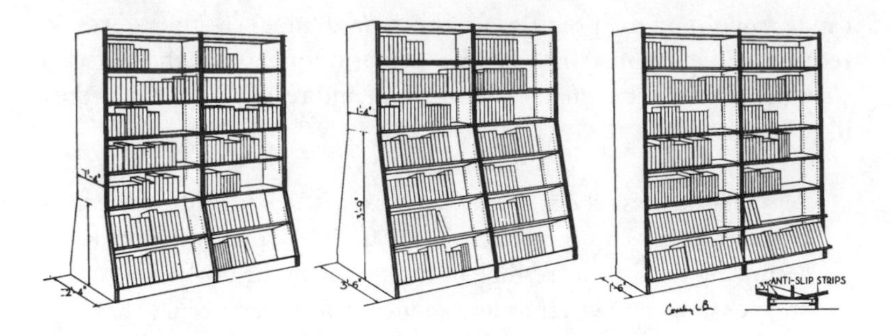

Bookcases with their lower shelves tilted upward have been recommended for libraries so that patrons could more easily identify books. Such shelves, which are also made more stable by their broader base, have come to be common in bookstores.

slight tilt is not enough for easy reading of the titles of books on the lowest shelves, however, and so the bottom shelf or two is often tilted even more or brought out further by the flare of the bookcase's base so the browser can read the titles without having to move out of arm's reach of the books. Given the narrowness of the aisles in some bookstores, the reader could not even get far enough away from a conventional bottom shelf to see what it held. Stooping down to look could work for the supple, but that tends to block the aisle for another customer, which merchants wish to avoid.

Though the principal purpose may be to bring the books into better view, this flaring of the bottom of a bookcase also gives it stability, as the sloped legs do the Eiffel Tower. Similar bookshelf arrangements were talked about for libraries as early as 1940, as noted in *The American Public Library Building:*

> Due to the reluctance of many readers to lean over to examine books on the lower shelves (oftentimes influenced by poor eyesight) several libraries have gone to the trouble and expense of raising the base of the bookcase and tilting the two lower shelves to make the backs of the books more legible. One of the chief objections to doing this is that the base of such a bookcase often occupies from 10% to 20% more floor space and thus narrows the aisles. The loss of book space within the case itself is also high,

while the cost of construction is considerably increased. However, in circulating rooms where crowds of people look over books and it is important to make books as attractive and convenient as possible, sloping shelves are well justified if sufficient space is in any way available. With the pressing necessity for making more books accessible to the public, it is not likely that sloping shelves with wide-based cases will become prevalent. On the other hand, it is significant that the equipment companies are now selling quantities of ordinary shelving with the two lower shelves brought out 2″ and tilted, a sensible compromise. Sloping shelves should have pebbled or corrugated rubber strips on the surface to keep books from sliding.

Not all library or store shelves are uniform in appearance and arrangement, and this is usually nowhere more evident than in older independent bookstores that grew as business grew, or in used bookshops whose shabbiness is more affected than their prices warrant. The prices are reasonable in the Book Exchange, located in downtown Durham, North Carolina, which long before the age of superstores advertised itself as "The South's Greatest Bookstore." The Book Ex, which deals in everything from new to used and from trade books to textbooks, must have begun as a small storefront operation, but as its stock began to accumulate, there certainly would have been more and more need for additional bookshelves. And so they were built in a bewildering range of styles, and mostly left unpainted. When such a store buys a library from an estate, the bookcases are often included, and thus many odd, unmatched cases are incorporated here and there into the Book Ex as they are in so many used bookstores.

The construction of bookshelves, whether for a store or study, can seem as Sisyphean a task as mowing the grass. I once built a wall of shelves into the living room of an older house, and I wondered if I would ever finish. Even though I knew rationally that I needed only a finite amount of lumber, which I purchased in the appropriate lengths after measuring and remeasuring the dimensions of the wall, and even though I saw the pile of lumber shrink as the bookshelves grew toward the ceiling, there seemed to be an unrealness to the process. It was, unlike turning the pages of a book, too much of the same thing, day after day. The plot did not develop as it does in reading, and my measuring, sawing, leveling, hammering became mantras to oblivion.

When the bookshelves were finally finished, the task of painting them had to be faced. This proved to be even more Sisyphean, for the boards that had only one significant dimension—length—suddenly had multiple surfaces that seemed to be elusive against the wall whose color they were to adopt. In time this task too became completed, but I often think of it when I encounter an expanse of empty bookshelves, whether painted or bare. If painted, I think of myself painting; if unpainted, I envy the merchant who insisted on getting books rather than paint on the newly constructed shelves.

On a later occasion my wife and I had a study built, including the bookshelves and cabinets that were to line its long windowless wall. We wanted the shelves to be adjustable, and so the cabinetmaker ran two strips of plastic shelf supports up each side of the vertical boards defining the bays of the bookcase. Neither my wife nor I had ever seen such a system, in which the plastic had projecting shelf supports every inch or so. The supports were wedge-shaped, and they were formed, as so many plastic articles are, with an integral hinge that allowed the supports to fold back into the plastic strip so that the shelves could be moved up or down without having to be removed from the bookcase.

One of the few ways a bookshelf can come to grief is by slipping off its supports or by having the supports themselves fail. We were familiar with plastic peg devices that had broken in some of the shelves we had in another room, and so we questioned the durability of the new plastic supports, but the contractor assured us that they were the latest thing and were much more durable. They did seem to work fine for a while, but soon some of the supports broke at the hinge, and the whole strip began to separate as the small nails that held it pulled out of the soft wood that had been used in the bookcase. We had, apparently, adjusted the heights of the shelves more often than the supports could take, and we had loaded the shelves with more books than the supports could handle. As the shelves became stuffed with books, however, and we ceased moving them about, the plastic strips came to be hidden behind books, and supported by them, and forgotten.

I had never even considered building these shelves myself, but in the interests of economy we had decided to paint the room—including the bookshelves—ourselves. The walls and woodwork were a relatively easy job, especially with a paint roller, but the bookshelves again proved to have more space for paint than they would for books. For some reason, rather than paint the bookshelves we stained them, perhaps because the wood looked so fresh and clean or because the shelves were supported by brown plastic strips. For whatever reason, staining seemed to be the appropriate and easier alternative, but it proved to be the more difficult because the walls behind the bookshelves were to remain painted. In time this task too was finished, and it was almost immediately clear that we would have liked the shelves better painted. We could not face the prospect of redoing what we had already done, however, and the darkly stained shelves remained a reminder of that in an otherwise bright study.

Not all studies or bookshops need even present the prospect of their bookshelves being painted or stained. Among the latest shops are those on the Internet, of course, and whether new or used these tend to have no shelves that the customer will ever see. The convenience of these virtual bookshops with virtual bookshelves can be enormous, their titles seemingly countless, and their prices attractive. However, without being able to browse among honest-to-goodness shelving, whether home- or factory-made, shopping in such a metaphorical bookstore can seem to be more like using a library catalog, and a computerized one at that, than visiting a bookshop. But to those who have experienced the pleasures of getting a hard-to-find book delivered a day after ordering it, these new stores may appear to have poetic overtones. As the poet Marianne Moore knew imaginary gardens with real toads in them, we now can know imaginary shelves with real books on them.

If some computer scientists and engineers succeed in their dreams, the future of the book itself will be such that bookshelves in bookstores, libraries, and homes could be a thing of the past. At the Media Laboratory at the Massachusetts Institute of Technology, a research team has been working on what it terms "the last book." This volume, known as "Overbook," would be printed in electronic ink known as e-ink, a concept in which page-like displays consist of microscopic spheres embedded within a matrix of extremely thin wires. The ink particles, which have one hemisphere black and one white, can be individually flipped by a current in the wire to form a "printed" page of any book that has been scanned into the system. According to its developers, the last book could ultimately hold the entire Library of Congress, which is of the order of 20 million volumes. The book one wished to read would be selected by pushing some buttons on the spine of the e-book, and the display on its e-inked pages would be rearranged. In time, the developers of this twenty-first-century technology claim, such books could also incorporate video clips to give us illuminated books that were also animated.

While Overbook may still be under development, several forms of electronic book were being promised for the 1998 Christmas season, or soon thereafter. With names like Rocketbook—Rocket e-Book—Softbook, and Dedicated Reader, they have capacities ranging from four thousand to a half million pages of text, downloaded from the Internet, and come in formats that look reminiscent of Etch A Sketch pads that display one page at a time or, in the case of Dedicated Reader, like a conventional book opened to two facing pages. A fourth book, occasionally named the Millennium Reader, weighed in at less than a pound and under $200, thus signaling that a new competitive era in bookselling had begun. Early readers of these e-books have found them user-friendly and appealing, but whether they will commercially displace real books remains to be seen.

The bookstore in the technological environment that will produce more and more electronic reading products will also likely carry books on microdisks, which would be able to be displayed on microshelves. Unfortunately, as the sale of audiotapes, compact discs, and computer software has demonstrated, such small items need to be displayed in much larger packages lest they fall off the shelf and into the pockets of shoppers. In the final analysis, the disk to set a book in e-ink may have to be packaged in something the size of a conventional book, thus requiring a display case remarkably similar to the bookshelves we now find in bookstores. As much as this may be the future, let us return to the past.

CHAPTER NINE

Bookstack Engineering

A stack is any collection of things piled one upon another, such as a stack of pancakes, bricks, or books. There are also objects that take their name from the word: haystacks, smokestacks, bookstacks. The idea for bookstacks arose with the ever-growing need to find room to store the increasing numbers of books that the collections of libraries comprised. Systems of lecterns, stalls, and wall cases were successively simply too wasteful of space to house books that might be consulted only a few times a year, if even that frequently. In the nineteenth century the idea arose of keeping a library's collection of books in a space separate from the reading room, and this led to the development of the bookstack as we know it today.

My own most vivid recollection of a bookstack whose structure was apparent is in the mathematics library at the University of Illinois at Urbana-Champaign. The library is housed in Altgeld Hall, one of the oldest buildings on campus. When I used this library extensively in the mid-1960s, the reference and periodicals room was fitted with wall shelves around its periphery, which were supplemented with low bookcases reminiscent of standing lecterns parallel to some of the walls. The library's collection was housed in stacks behind the circulation desk, and like old library stacks generally these consisted of floor-to-ceiling bookcases that were lined up in ranges between which there was only enough room to walk. The problem of lighting such a confined space, the outer bookshelves of which might stand between windows and inner shelves, thus obstructing light from a good many of them, was notorious in the days before electric lighting. In Altgeld Hall there was in clear evidence the ingenious solution of the modular bookstack, in which the cast-iron and steel framework that comprised the bookshelves also supported the floors, which were translucent. The floors were constructed in segments, each no larger than a large modern ceiling tile, made of thick glass that looked not unlike that used in glass bricks. The glass transmitted and diffused enough light so that even the lower levels, where there was little direct illumination from windows, got enough natural light so that a book could be located. At the same time, each glass floor, which was also a ceiling for the level below, was thick and wavy enough to make objects above or below unclear. (This feature enabled women to wear skirts and dresses into the stacks without worry.)

With the introduction of electric lighting, such elaborate measures as glass floors were not so necessary to admit light but not licentiousness into the bookstacks. Reading the views of one librarian, writing in 1916, we can see why glass floors became all but extinct:

> Since natural light is no longer necessary in many stacks, why continue to use glass floors in book stacks? In an artificially lighted stack, they are without excuse. In the first place, glass floors are exceedingly noisy and few students at work in the stack can withstand this annoyance. Also, they are so slippery as to be unsafe, and frequently cork carpet or rubber mats have to be placed at the top of steps and at other dangerous places on the glass floors. If sunlight reaches these glass floors in the stack, the reflected glare is most trying. Then too, they crack either through expansion or contraction, or else because a metal shelf has fallen on them. Another disadvantage of glass floors, in a dry climate at least, comes from the putty used along the joints and edges. This soon dries out and is loosened by the constant walking on the floor. The result is that not only are the books and floors constantly powdered with putty dust, but small putty fragments have a penchant for dropping down the necks and into the eyes of readers and workers in the stacks.

A decade later, marble, which would "reflect considerable light" compared to glass, whose "value as a light reflector is much less than was anticipated," had become "in rather high favor." With the introduction of electric lighting, floors could be opaque, of course, and reinforced concrete, which was still a relatively new structural material, could be used. It was employed in the construction of the John Crerar Library in Chicago, a city of structural innovation, and it was recommended to librarians for its "durability and economy in construction." The matter of light transmission and reflection had taken a backseat to matters of cost and convenience, something that was not possible in the nineteenth century.

The idea of keeping books in stacks out of sight of general library patrons arose in Italy and Germany as early as 1816, but it was in the Bibliothèque Sainte-Geneviève in Paris that the concept of a closed bookstack was first fully realized. Since the library building was completed in 1843, well before electric light was available, the reading room was placed on the top floor, where it could receive the greatest sunlight. The great space, which could seat six hundred readers, was enclosed by a barrel vault supported on cast-iron columns. It was underneath this reading room that the bulk of the books were stored, in tall wooden bookcases that ran across the 53-foot width of the building, with the necessary aisles left for access. The bookcases were spaced 14 feet apart, perhaps as much to allow natural light to illuminate them as to provide room for the ladder that was necessary to reach the upper shelves.

When the library of King George III was acquired by the British Museum in 1823, it undertook to build a special room to house the collection. The King's Library, which measured 300 feet long and had shelves lining the walls from floor to ceiling, interrupted by a gallery, has been described as "perhaps the last great flourish of library design" employing the wall system. This was only part of the British Museum's collection of books, however, for the revered institution then also housed the national library of Britain. As such, there were rooms of books of a more common kind and rooms where people of a more common kind could sit and read and work at scholarship. Indeed, the Act of Parliament that created the British Museum in 1753 and provided for free admission to "studious and curious persons" was altered in 1810 to grant general entrance to "persons of decent appearance."

The museum's first reading room was a "narrow, dark, cold and damp" space that contained a single table surrounded by twenty chairs, which were sufficient to serve the needs of the fewer than ten readers a day, not all of whom consulted even one volume in the library of fifty thousand. As the collection and its use grew during the course of its first century, there was a succession of six ever-larger reading rooms. By the middle of the nineteenth century, even the latest British Museum reading rooms were inadequate, however, and the perennial problem of book storage persisted. In 1852, the principal librarian of the museum, Antonio Panizzi, sketched out a plan for the seventh reading room, which would become *the* Reading Room.

Panizzi's plan, which employed the latest methods of the "engineering age," was a brilliant use of underutilized space. He proposed to put an enormous circular reading room in a large interior quadrangle or courtyard, whose cornice "mouldings are particularly fine." However attractive, the courtyard "could not be used as a garden" because "the air did not circulate and the grass never looked very green as the surrounding buildings excluded the necessary sunlight." That such a gloomy space would remain unused by museum-goers and unpopular among staff is perfectly clear to anyone who has visited London on a sunny summer day, when every square inch of lawn in the smallest park in the smallest square appears to be covered with sunbathers who might be taken for sun-worshippers. Thus, Panizzi's suggestion that a new reading room be constructed in the museum courtyard met with little opposition.

The nature of the construction, which began in 1854 and was completed in 1857, was state-of-the-art, and Panizzi no doubt had been inspired by the success of the Crystal Palace that had been erected in Hyde Park for the Great Exhibition of 1851. Like the Crystal Palace, Panizzi's structure employed cast iron as a principal material. The circular reading room was more or less centered in the enormous rectangular courtyard, which was 313 feet by 235 feet, but did not fill it completely, so that the windows in the existing museum wings were not blocked. The 27- to 30-foot open space left between the museum and the reading-room structure was also said "to reduce the risk of fire spreading from one building to another." Surrounding the reading room in the new structure were multilevel bookstacks, ranging in total height from 24 feet to 32 feet, with the larger dimension being of the four-level stacks that ran around the outside of the reading room proper. The books were thus in close proximity to where they were to be used. The bookstacks, like the Crystal Palace, were constructed with a largely glass roof and formed a vast ironwork structure that supported itself and the books it contained. Indeed,

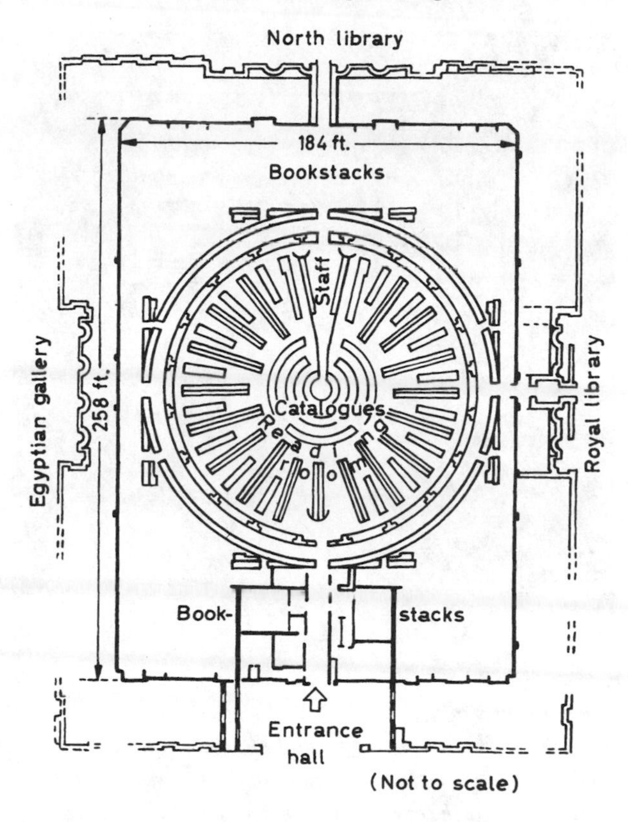

The British Museum Reading Room with surrounding bookstacks was built in an inner courtyard. A space of 27 to 30 feet was left between this Iron Library and the main structure of the museum so as not to block existing windows and to serve as a fire break between the structures.

Panizzi's building-within-a-building came to be known as the Iron Library.

Unlike in Paris, however, the bookshelves in the Iron Library were not all on one floor but were arranged in levels. The ranges of shelves were 7 feet apart, and light from the skylights could reach the lower levels through grate-like floors. Because such a large volume of space was available for storage and, for the time, utilized efficiently vertically if not horizontally, it was possible to store an enormous number

The bookstacks of the British Museum Library had grated floors and wide aisles to allow sufficient light from skylights to reach the lower stack levels. As shown in this photograph, in the late nineteenth century sliding bookcases were hung from the ceiling in the extravagant aisles to accommodate the overflow of books acquired by the library.

of books. In fact, Panizzi's plan enabled about 1.5 million volumes to be housed in the stacks, which by their cast-iron construction were considered largely fireproof. However, when an incendiary bomb fell through the glass roof during World War II, the air draft through the grated floor fanned the flames and "turned the stack into a blast furnace."

The Reading Room itself is an enormous space: its dome, at 140 feet in diameter, is 28 feet larger than that of London's St. Paul's, a foot larger than the dome of St. Peter's in Rome, and only 2 feet smaller than that of that city's Pantheon. The new library space was praised as "a circular temple of marvelous dimensions, rich in blue, and white, and gold." The project also had its detractors, however, and the museum's keeper of the manuscripts, who was the Italian-born

The famous domed Reading Room remains a part of the British Museum.

Panizzi's "great rival," condemned the space as "perfectly unsuited to its purpose and an example of reckless extravagance occasioned through the undue influence of a foreigner."

Having cast-iron ribs, the Reading Room's segmented dome is lighter than the concrete one of the Pantheon. Furthermore, because the modern dome's pilasters are made of slender cast iron, they do not dominate the peripheral floor space the way the piers do in the Pantheon. In fact, the pilasters painted with trompe l'oeil books on bookshelves blend in so well with the real and functional shelves that encircle the Reading Room that the structural components are virtually invisible to the casual observer. This—coupled with the fact that the doors that allow passage to the staff between the stacks on the outside periphery of the Reading Room and its two-level gallery are also painted to appear to be full bookcases—gives the illusion of one's being literally surrounded by solid bands of books and by presses that also support the great dome. In addition to the twenty-four thousand reference books on the floor level, available to be consulted directly by the patrons, the walls of the Reading Room around the gallery levels hold forty thousand more books, reaching 24 feet above the floor level to where the springing of the dome is located. The distance to the top of the dome is 106 feet, and it contains a 40-foot lantern.

The sunlight that came through this great oculus and through the great windows around the base of the dome illuminated the room, but only until four o'clock in the afternoon during winter months—or earlier when a London fog descended—because for a long time there was no artificial illumination. The use of gas lighting was considered in 1861, but it was discouraged by the head of the London Fire Brigade. Electric lighting was first attempted in 1879, with poor results, but soon four enormous arc lights were emplaced with reasonable success, thus allowing the Reading Room to remain open after sunset. Incandescent lamps were installed over the readers' desks in 1893, and similar lighting of the reference shelves soon followed. The installation of electricity not only allowed longer reading hours but also illuminated the stacks.

There was attention paid to more than books in the room's design. To deal with the cold in the winter, air was circulated around warm-water pipes in the space below the room and was thus heated before it flowed up through the hollow cast-iron framework of the readers' desks and, eventually, out through vents above the windows and around the dome's glass skylight. Further warmth was provided in the form of footrests beneath the readers' desks through which warm water circulated. The arrangement of the desks gave the library staff, who stood on the elevated platform in the center of the room, a view of patrons at the readers' desks, to which books were delivered from the stacks. (The floor-level architectural focus of the room was the circular segments of catalog desks, which served as buffets off of which "a formal champagne breakfast [was] ceremonially eaten" at the room's opening on May 5, 1857.) The radial arrangement of desks also provided a maximum of exposure and intimidation of those patrons who might deface books by removing leaves. It was at one of these desks that Karl Marx worked for twenty years, and it is his desk that visitors in small tour groups have most wanted to have pointed out to them.

For all the brilliance of Panizzi's design, in time the Reading Room and its surrounding stacks were inadequate for the number of readers and books used and stored in the British Museum. There was an attempt in 1920 to expand the capacity of the stacks by adding a fourth story in some areas, but it strained the original structure. The very fact that the Reading Room occupied an interior courtyard militated against its expansion radially or laterally. The library could, of course, expand by displacing other departments of the museum, but the museum itself was growing in the number of artifacts it held and the number of visitors that came to see them. Some relief had been achieved as early as 1887 by adding movable shelves in the bookstacks, which were supported from above by rollers running along ceiling tracks. This was possible because of the extravagantly wide 7-foot aisles in the original stacks, but it was inconvenient because entire bookcases had to be moved out of the way to get at the books in those behind. The hanging bookcases, moreover, ultimately proved too heavy for the original iron structure, and in time two hundred fifty of them had to be removed.

"Man builds no structure which outlives a book," wrote the Victorian poet Eugene Fitch Ware, who penned his verse under the pseudonym of Ironquill, and his observation has perhaps never been truer than in the case of the British Museum's stacks. In the early 1960s a move of the library to another location began to be contemplated, and in 1973 a new entity was created in the British Library, with the promise of thus emphatically distinguishing with its own freestanding building the book collection from other museum artifacts. The cornerstone for a new British Library building across from London's St. Pancras Station was laid in 1982, and the movement of twelve million volumes into the new library was to be completed in 1999. The British Museum Reading Room was to be renovated, stocked with a reference collection, and opened to all museum visitors "to celebrate the new millennium" in 2000, thus preserving what is to many a book- and library lover a hallowed space. The stacks surrounding the Reading Room will, however, give way to a glass-enclosed space promoted as "London's first covered public square," to be known as the Great Court.

By the fall of 1998, when I visited London, the great dome was beneath the shadow of a tower crane with a reach "longer than any other crane currently in use in the UK" and capable of lifting materials in and out of the courtyard over the building's main entrance on Great Russell Street. The "redundant bookstacks" had already been destroyed, in preparation for the space between the Reading Room's dome and the perimeter of the museum's courtyard that Panizzi captured for books and readers to be modernized and roofed in—by an enclosure that has been likened to I. M. Pei's pyramidal entrance to the Louvre. The new glass roof will open up the periphery but will not let any more sun into the courtyard. How ironic, though, that the space vacated by books will house, among other things, a terrace restaurant that will likely be better illuminated by natural light than the old stacks ever were.

It is also depressing that so many of the desks in the reading rooms in the new British Library building are arranged beneath ceilings barely higher than those in the old iron bookstacks. The most striking part of the new building, besides its entrance court and its deference to the towers and turrets of St. Pancras Station, may be the books of the King's Library removed from the British Museum and entombed in compact shelving within a glass core reminiscent of that at the Beinecke Library. Unfortunately, unlike at Yale, where the rare-book collection is the building's raison d'être, in the British Library the entombed books serve as a backdrop for scholars and librarians taking tea in a stack of cafeterias. There is a tradition of libraries breaking tradition, however.

In 1876 in America, book storage space in Harvard College's library, housed in Gore Hall, had to be expanded. This was done in a radically new way by building a shell of an addition, piercing it with rows of small windows, and covering it with a roof that was supported by the masonry walls:

> Into this were packed book ranges, row on row, tier on tier, with only enough vacant space to give access to the books. The aisles between the ranges were 28 inches wide and the tiers 7 feet high, allowing the topmost of the seven rows of shelves to be easily reached. The stack was six tiers high, self supporting throughout and depended on the building for protection only. The vertical shelf supports were of cast iron open work, the deck framing of rolled wrought iron, the deck flooring of perforated cast iron slabs and the shelves of wood.

This "prototype of the modern bookstack" would enable the library to grow its collection for another few decades, but by the beginning of the twentieth century a new expansion of library facilities was necessary. This time it was decided to demolish Gore Hall and its bookstacks and build a new library, to be known as the Harry Elkins Widener Memorial Library, which was completed in 1915. The demolition of the masonry walls of Gore Hall revealed the principle of the independently supported bookstacks, for when the walls were down the shelf supports stood exposed as a vast three-dimensional gridwork, much like a giant jungle gym. The bookstacks had in fact supported the floors, rather than being supported by them.

By the time Widener was built, the technology of bookstacks had advanced to a considerable maturity. The new library was fitted with the latest in storage technology, which had been developed for the Library of Congress, the institution of which dates from 1800. For the major part of the nineteenth century, the Library of Congress was housed in the U.S. Capitol. With the Copyright Act of 1870, which required that books seeking copyright protection be sent to the Library of Congress, its store of books soon became overwhelming. A new building was authorized in 1886, and it was to be completed in 1897. The construction project, including the design and erection of the bookstacks, was under the direction of the American civil engineer Bernard R. Green.

Bernard Richardson Green was born in Malden, Massachusetts, in 1843, and graduated from Harvard's Lawrence Scientific School with a degree in civil engineering in 1863. He spent thirteen years working with U.S. Army engineers on the construction of fortifications in Maine, Massachusetts, and New Hampshire before moving to Washington, where he was responsible for the erection of large public structures, including the State, War, and Navy Department buildings, the Army Medical Museum and Library, and the Washington Monument. In addition to the traditional problems that had to be solved in constructing a building for the Library of Congress, Green faced the specialized task of designing the bookstacks, for which he invented a new solution.

The problem that faced Engineer Green was to improve on the system used at Gore Hall, whose wooden shelves he believed presented a fire hazard, collected dust, hampered air circulation, and were poorly lighted. He addressed these objections by devising a stack system that was entirely cast iron and steel, with metal shelves that were gridded or slotted to reduce weight and improve air flow, and

The civil engineer Bernard R. Green was the inventor of the bookstacks that were installed in the Library of Congress in the late nineteenth century.

with floors of glass or marble that allowed light to pass through or be reflected, thus brightening the storage space. Green conceived a bookstack that was strong, flexible, and user-friendly, and he had drawings and a full-size working model constructed. The job of supplying 43 miles of shelves to hold some two million books was let out for bids, and the lowest bidder was the Snead & Company Iron Works, which was then located in Louisville, Kentucky. The completed stacks were put into service around 1895. The new bookstack "was so great an improvement over all forms of book storage that had preceded it that the library profession accepted it forthwith as not only a better book storage, but as a definitively perfect one." Furthermore, "library buildings, and their stacks, had clearly at last assumed their final form," or so it seemed at the end of the nineteenth century and for some decades into the twentieth. Like all claims of perfection in technology, those for Green's bookstacks would in time be superseded by enumerations of the faults and shortcomings of the system. But at the turn of the century that kind of criticism was years off.

The design of what came to be known as Library of Congress stacks, or alternately the "Green (Snead) standard," resulted from Green's systematic approach to the problem as one in engineering. In fact, it was fundamentally a problem in structural engineering, with ancillary problems relating to fireproofing, ventilation, dust removal, and illumination, which also had to be solved in an optimal way, if at all possible.

The main structural demands on a bookstack of the kind Green designed are for the bookcases to support themselves, the books, and the floors on which people walk and move books within the stacks. To a structural engineer, a framework that satisfies all these requirements is relatively easy to design. The vertical members of the bookcases can serve as structural columns to support everything else: shelves, floors, and the columns above, which in turn buttress shelves, floors, etc., plus the books and people for which the stack is to function. Once it is clear how the columns are expected to function, the loads on them can be calculated. For example, the engineer might allow 10 pounds per cubic foot occupied by the ranges of bookshelves to account for the bookshelves themselves and their shelf underpinnings; 20 pounds per cubic foot for books; 15 to 25 pounds per square foot of floor area for the floor itself; and 40 pounds per square foot for the people that use the stacks. (Even though a normal adult weighs more than 40 pounds and can stand on less than a square foot of floor space, the floor section itself that supports a person's total weight is several square feet, at least, and it is through this section that the weight of the person is transmitted to the structural frame.) The weights of the shelves, books, and floors are collectively known as the "dead load" on the structure, because they are for all practical purposes unmoving. The weight of people that can move about on the floor and present different forces on different parts of the structure at different times is known as the "live load." (Today, the dead load might typically include a building's floors and structural columns alone, with the now moveable bookcases, books, and people constituting the live load, which might be taken at 150 pounds per square foot, floor area and not volume now being the common measure of useable space. It is this design load, as it is also known, that remains a characteristic of the structure and that often limits its future use. Libraries so designed cannot accommodate modern compact shelving, which typically requires a floor rated at about 250 pounds per square foot.)

Though the first engineering challenge to designing a bookstack might be the structural aspects of the problem—for if the stacks do not stand up, nothing much else matters good engineering demands that other functional aspects be considered also. One of these functional aspects is to let as much light into the stacks as possible.

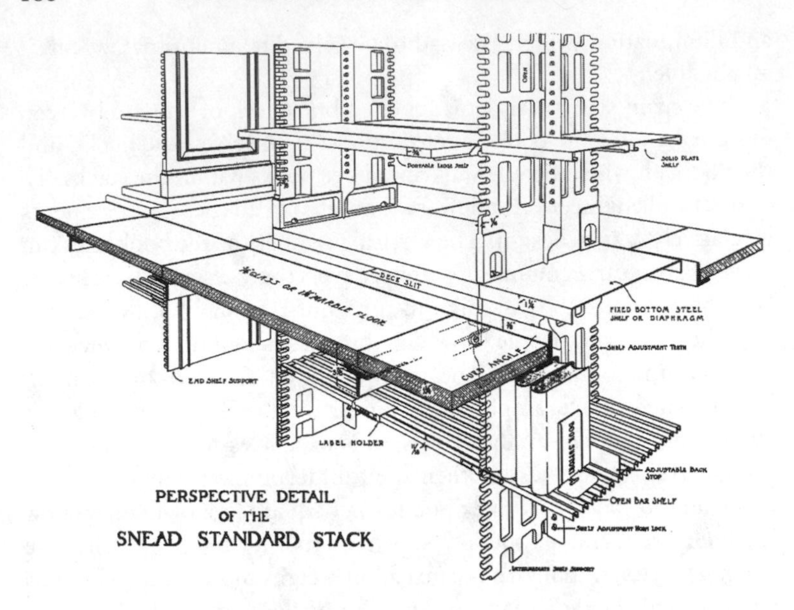

The structural elements of the Library of Congress bookstacks that support the shelves also support the floors on which staff members walk to retrieve books.

Another is to reduce the threat of fire as much as possible. The two are not unrelated.

Tightly closed and closely shelved books do not burn readily, but the shelves themselves and other wooden parts of older library structures, such as in Gore Hall, did and do. Hence open flames were used for illumination only as a last resort. But they were used, even though they evoked "great heat and more or less smoke" and they were "always difficult and inconvenient to control." Furthermore, the flames gave off a light whose "brilliance was limited and its color dull—in every respect greatly inferior to sunlight." By the end of the century, the incandescent bulb as a source of light was a viable alternative in bookstacks and elsewhere: "It is far more nearly white, radiates but moderate heat in proportion to candle power, may be placed with safety anywhere, even in one's pocket or mouth." (This last reference appears to be to a flashlight.)

But the design and construction of the Library of Congress took place before the electric light was commonplace, and, according to Green, writing about a decade afterward,

> Until very recently, in fact down to the time of the incandescent lamp a few years ago, daylight was almost wholly depended on for finding books on the shelves. Consequently the prime effort in design and arrangement was to get daylight into the shelf spaces through windows and skylights, using the ground space and special locations on the lot indispensable to that purpose. This was done with special pains in building the Library of Congress. . . .
>
> Hitherto, book stacks have generally been placed at the outer wall lines of the buildings and wide open spaces left around them to admit as much daylight as possible. Skylights have been provided in the roofs and clear light wells within the shelf rooms for the penetration of the light down and between the shelf ranges. As much or more space on the lot was given up to the admission of daylight as to the shelving and its communicating spaces combined.

Green went on to note, however, that daylight was "the most unequal and unsteady of all human dependencies, under the ever-changing position of the sun and condition of the weather." Furthermore, bright sunlight is the enemy of books—"books, in fact, are much better off in the dark"—and so he noted that "when we make anxious provision to let it in, we must make similar expensive provision for keeping it out." Since the Library of Congress stacks had six hundred windows, raising and lowering the shades would have been a very time-consuming task had they not been fitted with special mechanical controls so that a single attendant could operate one hundred fifty shades at one time. It was this kind of extraordinary effort that led stack designers to rely less and less on natural light, and hence less and less on windows and skylights themselves. The use of electric lights in bookstacks became firmly established in the early part of the twentieth century.

With the electrification of libraries, their stacks not only could be arranged without regard to natural light sources but they also, like the British Museum Reading Room, could remain open after dark.

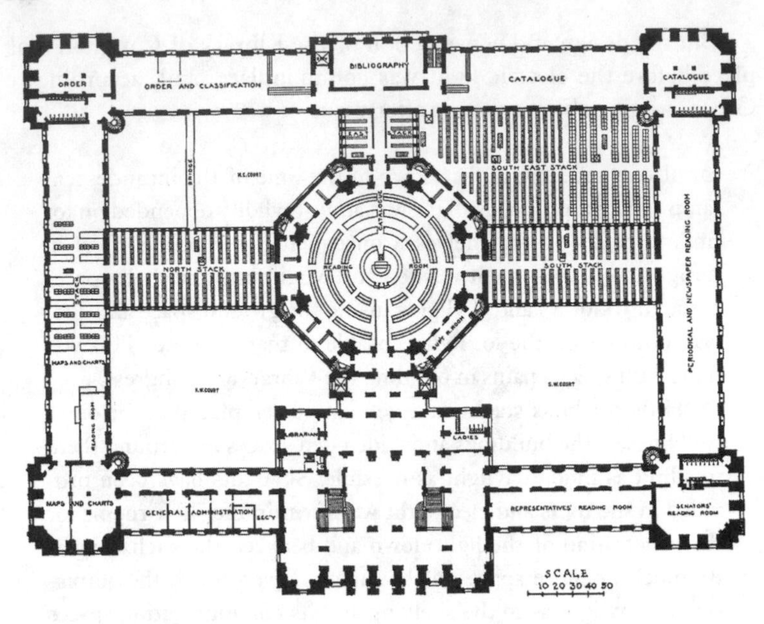

The first-floor plan of the Library of Congress shows the location of the north, south, and southeast courtyard bookstacks.

Thomas Edison, who made such a significant contribution to this advance in technology, had an elaborate library of his own at his research laboratory in West Orange, New Jersey. Built in 1886, when incandescent lighting was not yet taken for granted as a substitute for natural light, even in Edison's establishment, the library is a handsomely paneled room with many large windows. The shelves are arranged in a modified stall system, each bookcase shouldering some of the load of the wide galleries above. On the gallery levels, the cases also are arranged more or less according to the stall system, thus providing much more capacity than would a wall-system arrangement. The size of the windows allows a good deal of light into what would otherwise be dark nooks. The windows were also fitted with shades, however, perhaps in part so that the legendary inventor could catch one of his famous catnaps in a darkened alcove.

The problem of letting light into a library building while at the same time minimizing the harmful effect it has on the books stored there has been admirably dealt with in more recent years in the struc-

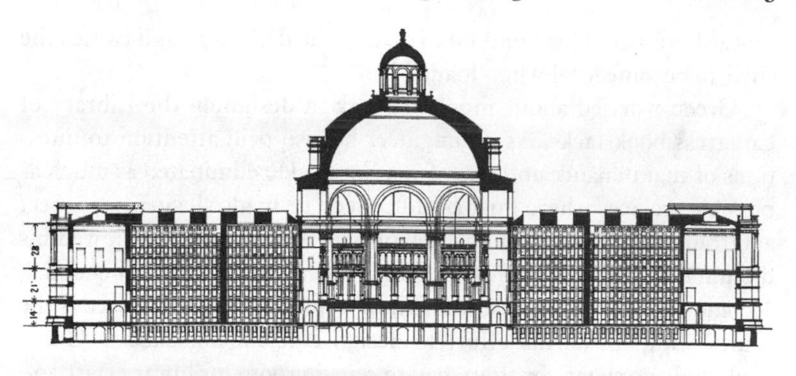

This section view of the Library of Congress shows the main reading room flanked by the north and south stacks. The multitude of windows letting light into the stacks overlook the northwest and southwest interior courtyards of the building.

ture that houses the Beinecke Rare Book and Manuscript Library at Yale University. Designed by Gordon Bunshaft, of the firm of Skidmore, Owings & Merrill, the upper and most visible part of the building has no windows in its Vermont marble and granite walls; but the large translucent marble panes of the structure are as little as ¼ inch thick, and they allow enough light to pass through them to illuminate the space within.

As the architect Le Corbusier would see a building as a machine in which to live, so the engineer Bernard Green saw the bookstacks that he devised as "not a building but a piece of furniture which may be set up and stand alone in any room adapted for its reception." He allowed for air circulation by leaving slits beneath the ranges of bookshelves; he devised uprights of open cast-iron work and shelves with slits to further facilitate air circulation.

The strength of Green's "open bar shelves" was confirmed by the Engineering Department of Columbia University, where tests on 10-inch-wide shelves showed "a shelf 3 feet 6 inches long with [¾-inch] deep bars to be stronger than a shallow [⅝ inch] bar shelf only 3 feet in length." Even the strongest shelves sag or bow under the books they carry, however, and Green dealt with this problem in the same way it is addressed in especially long and heavy steel and concrete girders: "When shelves are to be *extra heavily* loaded, they are given a slight upward camber. This bend takes care of the deflection and causes the shelf to become level when loaded."

Green worried about more than sag in designing the Library of Congress bookstacks. As an engineer he also paid attention to questions of maintenance and user-friendliness. He eliminated as much as possible crevices where dust could collect; he made all surfaces reflect and transmit as much light as practicable. And finally, he followed his dictum that "that scheme of shelving which, other things being equal, accommodates the greatest number of volumes in a given space is the thing." But, for all his effort to design efficient bookstacks, Green could only go so far, for there had to remain room for library staff and patrons to move among the books to shelve and retrieve them. Indeed, the conventional stack could evolve only to the point where, on average, aisles took up 65 percent of the floor space, leaving only 35 percent available for tiers of bookshelves. The problem posed by that limitation would not be dealt with truly effectively for years to come.

The British Museum Reading Room and some of the Library of Congress stacks were, of course, constructed in the courtyard and interior spaces of their buildings, thus presenting no egregious conflict with architectural treatment of facades. This was not the case with the New York Public Library, completed in 1910 on the site of the old city reservoir and designed from the outset to incorporate stacks with over sixty-three miles of shelving capable of holding over three million volumes. In what became a familiar arrangement, the stacks are underneath and support the floor of that library's great reading room, to which the books were "simply drawn directly up through the floor from the mine of human knowledge beneath." The idea of putting the reading room on a higher rather than a lower floor in the building was reminiscent of arrangements as old as medieval libraries. The closely spaced narrow slit windows in the west-facing facade of the New York Public Library that looks onto Bryant Park are also reminiscent of the spacing in a medieval library fitted with lecterns or bookstalls. The marble walls between the windows support in an almost column-like way the large arched windows through which more intense light reaches the reading room. Indeed, it is because so many library buildings have such telltale window arrangements that it is easy to identify where their stacks are located. For example, walking down York Street in New Haven, one can see such a

The rear of the New York Public Library building, overlooking Bryant Park, displays the lines of narrow windows that are characteristic of a bookstack area. The much larger arched windows indicate the location of the main reading room above the stacks.

window arrangement in the tower of Yale's Sterling Memorial Library, thus marking it as a stack tower.

Writing in 1933 of the library of the future, Angus Snead Macdonald, president of what had become the bookstack manufacturer Snead & Company, described an imaginary tour through such a library conducted by its director, who considered it as "a working laboratory for all kinds of people rather than as a monumental reading place for the comparatively few congenital 'book worms.' " According to the director, "absolute control of air conditions, temperature, humidity, and dust-content should be accepted as a matter of course" in such a "People's University." He went on to describe "our nondependence on window ventilation and light," and went on further to explain:

> We have windows, it is true, but we depend on them only as a means of looking out of doors. They are never opened. Consequently we do not have to make our rooms unnecessarily high so as to allow the daylight to penetrate to the opposite wall. . . . We

feel that daylight as well as window ventilation is much too fickle and costly to receive consideration in working up a library plan.

When the Bodleian Library constructed a new building, which was completed around 1940 in the midst of war, it was built with a steel frame and concrete floors. Furthermore, with electric lighting well established—if not taken for granted—by this time, there was no need to worry about large windows or glass floors, and in fact three of the floors were built below ground level. (Such basement and sub-basement storage of books has become common, and entire libraries have been built underground. Such was the design for the new undergraduate library begun at the University of Illinois in the late 1960s. The plan used the quadrangle space in front of the old library while largely preserving the quadrangle vista.)

Atop the three basement floors of the new Bodleian, there are eight above-ground floors of books, and the bookstack core rises 78 feet above street level. The Bodleian stacks rise from the middle of a neoclassical building that is significantly lower than the stack at its core. However, since the core is set back from the exterior of the building proper, it is hardly noticeable from the sidewalk. Indeed, one has to stand on the top steps leading to the Clarendon Building across the street even to see that there is a stack core.

In contrast, the bookstacks added in stages behind the main library at the University of Illinois dominate street views of the building from its back, and they look more prison- than library-like, for whereas medieval libraries were designed with windows to let light in, many a modern library stack has been built, as Macdonald predicted, with anywhere from many small to few if any windows for admitting light. Those windows that are incorporated in the design appear to be employed more for psychological than for physical reasons. Such considerations might not have been adopted at all had only staff been allowed in the stacks, but beginning in the last decade of the nineteenth century there were increasing calls to open bookstacks to patrons, some of whom were no doubt claustrophobic.

Many modern stacks did in time become open to all library patrons, but library staff still had to find their way about those stacks that remained dark and closed. One student worker, who after the Second World War was attending the University of Iowa on the GI

The new Bodleian Library, completed during the Second World War, has its stack area located in the central core of the building. Although its top levels can be seen in this photograph taken from an elevated vantage point, the stacks are not readily visible from the sidewalks surrounding the building.

Bill, recalled vividly how he and his fellow workers retrieved books from a then-unfinished floor in the new Main Library building:

> The near Stygian darkness was unbroken except for a few electric lines with naked 60-watt bulbs every 20 feet or so, hanging over every other aisle. All call slips came over tubes [and] once we got the request, we picked up our sturdy two-cell flashlights and ducked into the shadowed alcoves where we beamed each shelf until the volume was illuminated and—most of the time—retrieved.

The use of flashlights was hardly anything new; as we have seen, Green reported their being kept in the pocket and held in the mouth around the turn of the century, and "hand-lamps" had been adopted as early as 1912 in the British Museum Library. These made it possible to provide books for patrons as late as 4:30 on winter afternoons and up until a half hour before closing time in summer.

Dark or not, and claustrophobes in them or no, many modern bookstacks do have low ceilings. This keeps the overall external height of the stack as unobtrusive as possible, and allows the top shelves to be reached with a step stool or small two-step ladder. However, as Dewey noted, since "sum people ar dizzy or unstedy on their feet when raised only two or three steps from the floor," it was "an old library device to hav a wood upright, 3 to 5 cm in diameter, extending about 100 cm above the step." In the absence of stools or ladders, but perhaps only for those without a propensity toward dizziness, sometimes "a bracket step" upon which to place a foot, and a handle attached somewhat higher on the bookcase, were provided whereby patrons could, as if boarding a streetcar, pull themselves to the necessary height and so retrieve with their free hand a book desired. Such a flourish added to bookcases was worthy of the Victorians who did it.

Stacks could never be too low for some library staff, such as the unusually small man who worked as a reference librarian at the research laboratory where I once was employed. He mastered his limitation by taking patrons in tow so that books on the upper shelves were not, metaphorically speaking, out of reach. That library had been built decades before regulators were sensitized to the needs of the disabled. Such a disregard for them was in sharp contrast to regulations that defined much in the new engineering library that opened at Duke in the early 1980s. It had its shelves spaced exceptionally wide apart, as then required by law, I was told, so that wheelchairs could move between them with ease. The considerate regulation that mandated the spacing was confusing to me, however, since it did not also restrict the height of the shelves. Perhaps the wheelchair-bound patron was expected to take a library staff member in tow to reach the upper shelves. A further irony of the arrangement of the shelves was that they were on the uppermost floor of the library, which was accessible from the entrance level only by stairs. To use the elevator located elsewhere in the building, a patron had to leave the library and reenter it through an emergency exit on the upper floor. Such an unorthodox entry required the escort of a staff member, and so the disabled person could not browse through the stacks unassisted, as would seem to have been the intent of the regulations.

Even when they are readily accessible, books can be hard to see. The ubiquitous availability of electric lighting in the later twentieth century has allowed libraries to install bookshelves with little attention as to how they are arranged with respect to natural lighting, and in some cases this seems to have led to a total disregard for lighting altogether, natural or artificial. The library remarkable for the generous aisles just described is located in a building that also houses the offices of the dean of engineering and some research laboratories, but it is the library that is the central feature of the building, the part that one sees straight ahead when one comes through the main entrance. Since the library occupies the back portion of the building, it has windows on only three walls. However, because of architectural decisions, only the northwest wall has a large number of windows—in fact a long string of windows architechtonically totally unrelated to the historical fenestration of libraries. Nevertheless, they do let in plenty of natural light, which makes the library pleasant to work in, especially if one is seated at a desk or open carrel beside these windows.

Yet the arrangement of some of the bookshelves in the library almost totally ignores the windows. On the top floor, where the monographs are housed, the bookshelves are arranged in what one might expect to be the only logical way. They are perpendicular to the main bank of windows, so that the natural light that flows through them illuminates each aisle to the maximum extent. On the main floor, however, where current periodicals are displayed like medieval books in cases with lectern-like shelves, the bookcases are arranged parallel to the windows. This provides plenty of natural light for the shelves facing the windows, but those on the other side of the bookcase are in shadow. There is fluorescent lighting in the ceiling, of course, and so one hardly notices the difference when the library is open and the lights are on. However, on those occasions when I have used it after hours on weekends or during vacation periods, when the lights were off, I have found the difference to be almost that between day and night. To read a magazine or journal found on the dark side of the case, I have had to carry it around to where the windows are.

On the lower floor of the library, where the bound periodicals are kept, on shelves that match in design those on the top floor, the bookshelves are also arranged parallel to the windows, which thus admit natural light to only one in a dozen aisles. Here the effect of natural lighting is even more striking between the shelves farther from the windows. When the lights are out, it is as if one is in a cavern, and it is not even possible to find a volume in the dark, much less read it. It is a clear example of the century-old but still sensible words of Melvil Dewey being ignored: "Aisles lined with books should as far as possible point directly towards the best light. Some stacks ar ruined by overlooking so obvious a law. If arranged across this line, the first book case cuts off all light."

In Duke's main library, the arrangement of the ranges of bookshelves has evolved still further from a consideration of the source of natural light. The square floor plan of the building, as geometrically distant as possible from the long, narrow standard of medieval libraries that had to rely exclusively on sunlight, enables a maximum number of shelves to be as far from what windows there are. The building was constructed in the late 1960s, when there was no question but that the library would rely on artificial light. In the stacks, where the ceiling is exposed concrete, the fluorescent lights are arranged in lines perpendicular to the aisles, a further example of how far from historical library architecture modern buildings have evolved.

The lighting on the main-floor reference area is integrated into a false ceiling, as dictated by the more formal nature of the space. The lights are naturally arranged according to a grid, with the long fluorescent tubes forming large square patterns framing smaller squares of light sources. It is all very regular geometrically, and the clear axes of the light system are parallel to the building's outside walls. When I first used this library, the shelves were set up parallel to one axis of the lights, an expected arrangement. Ideally, there should be a line of light fixtures centered over each aisle between bookcases, but for high-ceilinged rooms this is not so important, and the architectural treatment of the ceiling lights seemed to provide plenty of appropriate illumination.

At some time in the 1980s, however, new carpeting was installed on the main floor, and the occasion was taken to rearrange the bookcases. But instead of being kept on a grid that was parallel to the walls and to the lighting fixtures, the reference shelves, of which there are many, were turned at an angle, so that their arrangement bears no relationship to the geometry of the building or, perhaps more significantly, to the axes of the lighting. This is hardly noticeable to the library user who is looking at books rather than architecture, but it is an unsettling reminder of how we have forgotten the historical relationship of bookshelves to light.

In the mid-twentieth century it became the fashion in library architecture to design buildings as open-floored structures in which furniture, including bookcases, could be moved about at will. The Green/Snead Library of Congress bookstack that six decades earlier had been declared "perfect" was now viewed as disadvantageously locking a stack arrangement into the configuration of its construction. In the new approach, reinforced concrete floors carry the loads of bookshelves, so that they can be arranged without regard for window placements. This apparently has the appeal of flexibility in the light of indecision, for planners need not look at the functional and aesthetic requirements of their space and its fittings with any degree of finality; they can always change the use of the space as whim and fashion and consultants dictate. It is unfortunate that such has become the case, for it reflects not only a lack of sensitivity to the historical roots of libraries and their use but also rejects the eminently sensible approach to using natural light as a means of energy conservation if nothing else. There is little more pleasing experience in a library than to stand before a bookshelf illuminated not by fluorescent lights but by the diffused light of the sun. Direct sunlight can be an annoyance and have a downright blinding effect, of course, but it has been the challenge to architects and engineers since Vitruvius to orient their structures—and the bookshelves in them—to minimize such problems in institutional stacks and in private libraries alike. Let us hope that not all future librarians lose their heliotropic instincts nor lose sight of the bookshelves for the forest of bookcases in which they rest.

CHAPTER TEN

Shelves That Move

As every other prior system for storing books in libraries had achieved its maximum capacity and thus stimulated further developments in shelving, so the bookstack also, around the mid-twentieth century, reached its limits, and new solutions had to be found. Extensions of the stacks could be considered, of course, but especially where library buildings were crowded in among other buildings, this was not always an option. It was also becoming clear that even a new stack tower or an entire new building would provide only temporary relief. After perhaps another few decades at most there would be a new crisis.

Sometimes when space is not available, for whatever reason, resourceful librarians make room by encouraging the checking out of books and discouraging their return. I once worked in a research-and-development laboratory whose library rooms were overflowing with books. The librarian made no bones about the fact that she hoped that research staff members would each keep at least several dozen library books that were most pertinent to their work in their offices. If even half of the staff decided to return their books, they might not have fit physically into the library room, let alone onto the library's shelves. Some years later, when the Duke engineering library was still in its old cramped quarters, I began to notice another approach, in which seldom-used books were placed on top of cases, space normally meant to serve as a dust canopy over the books.

The anecdotal evidence of insufficient shelf space has been supported also by statistics. In 1944 Fremont Rider, the librarian of Wesleyan University, wrote of "the astonishing growth of our great research libraries" and presented hard data in support of his observation that for more than a century the collections of college and university libraries were doubling in size on average every sixteen years. To drive his point home, Rider projected out a century—to the year 2040—for the Yale University Library, which he chose as an example because the size of its collection and its growth were close to the mean of similar institutions. Rider noted that if it were to continue to grow at the historic rate, Yale's collection would swell from almost three million volumes in 1938 to approximately two hundred million in 2040. The shelving needed to house so many books would amount to six thousand miles. Furthermore, if a card catalog as it existed in 1938 were maintained to inventory all those books, three-quarters of a million drawers occupying eight acres of floor space would be required.

The card catalog of the Yale library today, having spread out from alcoves into the main traffic corridor, does occupy a good deal of floor space, but fortunately the introduction of computerized catalogs has tempered the acreage expected to be needed in the future. The rate of growth of libraries like Yale's has also slowed. Had the sixteen-year doubling time been maintained, by Rider's projections Yale would have had about forty million books by the year 2000. However, the rate of growth of the collection slowed in response to the realities of the economics of collecting and of shelving; in the mid-1990s Yale's library held just over ten million books, but this did not include such other items as newspapers, government documents, and manuscripts. Thus, while the situation might not be as bleak as Rider feared six decades before, there is still plenty of reason for concern. But librarians have been troubled about many things for a long time, and there has seldom been unanimity of opinion among them about problems or solutions.

As the end of the nineteenth century approached, William Poole, the librarian of the Newberry Library in Chicago, argued "why wood shelving is better than iron," citing cost, aesthetics, and other advantages. Wood was not only cheaper than the metal alternative but also "more tasteful and ornamental." Furthermore, wood was "a more amiable and benevolent material than iron, less harsh and abrading to the binding of books." Indeed, after the British Museum had installed iron shelves, they had to be covered in leather. So much was required "that it broke the market in Europe on a certain kind of kip," which is the undressed hide of young and small animals. Finally,

Poole ridiculed the choice of iron for its incombustibility. If that were a criterion of choice, he argued, "books also ought to be bound in sheet-iron, and some metallic substance, perhaps asbestos, substituted for paper." There were plenty of advocates of metal shelves, however, and their arguments were articulated by, among others, Bernard Green, who noted that wood shelves in wood cases "contrived to hold and hide indefinitely whatever got into them, especially dust, litter, and musty odors."

Just as debated as wood versus metal was the matter of "movable vs. fixt shelves," as Melvil Dewey declared in *Library Notes:*

> It has been an axiom of library economy that all shelves wer to be movable. A builder with proper machinery wil make them so at about the same price, and, as they need not be moved unless necessary, it seems a great advantage to be able to change. The objections ar: It adds something to expenses. Pins or some form of support must be provided, and often get lost or come out. The shelves warp much worse than if grooved firmly into the uprights. The shelves ar not as stif and strong as if built into a solid piece, where each shelf becomes the strongest kind of a fixt brace. To offset this, hevier stock and uprights must be used, and this adds to cost. Thicker uprights ar necessary also, to get room for pins. Except with the old-fashioned plan of sliding the shelf in a groove, there is nothing in the supports to prevent its tipping up if weight is thrown on the front edge. We hav seen many shelves of books thrown to the floor when readers stept on the edge of a low shelf or caught a high one with the hand. Bindings hav often been ruined by such falls. Finally, the uniform appearance of the room is much helpt by fixt shelving, in which the lines run with perfect regularity.

Many a librarian would echo Dewey's complaints and support his preference for long straight lines of shelves. Robert Henderson, "In Charge of Stacks" at the New York Public Library, wrote in the mid-1930s that "Rows upon rows of shelves, in unbroken line, especially when the books are in good order, have a classical austerity that is pleasing to the eye." Even when all the books had their spines straight up and at the edge of the shelf, however, the ragged line presented by their tops resembled not so much order as the graph of a random event like rainfall or the heights of librarians. To handle this, "sum recommended that cloth or lether falls be tackt on the edges of shelves" to "giv a certain finish to the shelf, and serv to even off the irregular heights of books, and to keep out the dust that would hav the fullest access to the tops."

Dewey also formulated a series of points of good practice regarding library shelving, which included the height of stacks and width of aisles. While he argued in preference of fixed shelving to gain a look of "perfect regularity," Dewey also recognized that when adjustable shelving was used it should be interchangeable. This meant that, in addition to shelves within the same section fitting anywhere within that section, shelves from adjacent areas—and from those on the other side of the room—should also be interchangeable. This is desirable so that the section of a book press that was initially fitted out with six shelves, say, could be fitted with a seventh of the same design and finish if the need arose. In too many libraries this could not be done, however, because "architects design these miserable shelves, builders put them in, trustees pay for them—sometimes almost double what the better ones would cost—and the poor librarian pays for the ignorance of all concerned." Dewey's complaint was fleshed out by Fremont Rider, as follows:

> Even in Wesleyan's relatively small library building our architect inflicted upon us no less than *thirty-seven* different lengths, sizes and styles of shelving, only a small minority of them of standard size and so interchangeable. In our stack alone there are seven different lengths of shelf, where there was no necessity whatever for more than one. (To multiply our annoyance four of these seven shelf lengths vary from each other by so little a difference that it cannot be detected by the eye, which means that these shelves have to be actually measured every time we have occasion to move one of them.) Such an absurd hodge-podge of uninterchangeability as this was, of course, expensive to install in the first place, just as it will be forever an expensive nuisance for us to use.

Even smaller public libraries, with adequate room when new or newly expanded, often "find themselves pressed for shelf space after five or ten years." Temporary relief can be gained by weeding out and discarding—perhaps at a book sale—volumes no longer popular and rearranging the collection that remains. But because tastes change, and because books of different kinds tend to come in different sizes, the reconfiguration of a collection often requires the adjustment of shelf heights here and there in the library. It is when this is attempted that librarians are frequently reminded of their frustration with architects and contractors.

Bookshelves are made adjustable precisely because not all books are of the same height and because of the need to rearrange them now and then. A common way of providing for the adjustment of wooden shelves in wooden cases is to have series of holes drilled in the uprights of the shelves so that pegs or pins can be inserted in them at the desired heights, an effective design in spite of Melvil Dewey's faultfinding. The holes are preferably spaced about an inch apart so that some degree of refinement in shelf height can be attained, but the holes obviously cannot be too large. And the smaller the hole, the greater the need for specially made pegs of the proper strength. It is imperative, of course, that the holes line up, but it is sometimes discovered only when the shelves of a new library are about to be stocked with books that the holes on one end of a shelf do not match exactly in height those on the other. There is the quick fix of shimming up the low end of the shelf, but every time shelves have to be moved the librarian is reminded of the problem. The precision achievable with moveable steel shelving installed in stacks beginning in the late nineteenth century did not present the same frustrations that wood-cased shelving did with keeping a long horizontal alignment.

By the 1940s, there was a sensitivity to the "psychological importance of cheerful colors" and this had led to their being baked on in the enamel finishes given to steel shelves, which had generally won out over wood. Such colors as "ivory white, light greens and grays, tans and buffs" whose "superior light reflecting properties" led to their replacing the "stock olive-green so widely used for office furniture," began to predominate in new shelving installations. The now familiar bracket shelf, which is supported at its back rather than along its ends, was at one time criticized for its being inconvenient, ugly, unstable, and uneconomical, but eventually it became familiar in libraries everywhere. This kind of shelf, which appears to have built-in bookends that double as book restraints and locations for the bracket hooks that fit into the slotted vertical structural elements that sometimes look too slender, is structurally a cantilever. The whole range of shelves supported in such a way in so many late-twentieth-century public and institutional libraries and additions can be made to move when one pushes a book back on the shelf, but the structural strength appears to be adequate, and the catastrophic accident is a rare event indeed.

There have been problems associated with industrial shelving used to support books and similar materials. One mishap occurred in 1968 at Northwestern University, where an empty section of "industrial shelving, freestanding, unanchored, and unbraced," which had just been moved, fell against some that were full of books: "A domino effect toppled twenty-seven ranges, spilling 264,000 volumes, splintering solid oak chairs, flattening steel footstools, shearing books in half, destroying or damaging more than 8,000 volumes." Unlike in E. M. Forster's novel *Howards End,* in which a falling man grabs onto a bookcase and causes himself to be buried under an avalanche of books, no one was injured in the remote stacks at Northwestern, but an employee was killed by the collapse of similar shelving that occurred in 1983 in the Records Storage Center of Ewing Township, New Jersey.

Earthquakes can cause otherwise stable ranges of bookshelves to tip over, and shelf sections should be bolted to walls braced transversely. This is often done by means of struts bolted across the tops of the shelf sections at regular intervals. The Huntington Library in San Marino, California, restrains the contents of its shelves with the kinds of bungee cords used to anchor articles to bicycles, motorcycles, and the like. When such precautions are not taken, libraries run the risk of having happen to them what occurred in 1983 to the Coalinga, California, District Library: "The card catalog toppled over, wall shelves collapsed, some stacks twisted, and two-thirds of the library's 60,000 books spilled to the floor."

For many years, librarians were more concerned with how many books could be stored on a shelf rather than the material of which the shelf was made, its reserve strength, or its stability in an earthquake. In a research library, where significant space cannot be gained by weeding out and discarding duplicate and outdated copies of books that no longer have a waiting list of readers, real bookshelf space has constantly to be found. Dewey, when he was librarian of Columbia College, described how that institution left bookshelves out of sections "to head hight wherever we wish an aisle." This provided convenient extra aisles through stack levels until the luxury of space fitted but unused for shelves could no longer be afforded.

Among the earliest solutions to adding shelf space to packed libraries was at Trinity College in Dublin, the Bodleian Library at Oxford, and the Free Library in Bradford, England, all of which employed the rolling or sliding bookcases that were installed in front of existing shelves. Whenever access to the blocked shelf was needed, the added case was rolled or slid away to gain that access. The task of getting books from the case behind the moveable one was not unlike going into the bottom of a tackle box fitted with hinged compartments or a toolbox with a lift-out tray.

The most widely publicized sliding presses were those installed in the late 1880s at the British Museum. According to that institution's librarian at the time, Richard Garnett, "the introduction of the principle at the Museum dates from the November evening of 1886, when, going down to attend a little festivity on occasion of the reopening of the Bethnal Green Library after renovation," he was shown its "supplementary presses." The first batch of new presses for the museum were ordered early the next year. Garnett described the new supplementary shelving as increasing the library's capacity while not requiring any new space. This was possible, according to Rider, "because the British Museum aisles involved had been inordinately wide to start with." Indeed, subsequent stacks did not have the extravagant 7-foot-wide range aisles of the museum.

Anticipating the idea of compact shelving, Dewey described the "suspension book cases" of the British Museum type with approval and noted that, with the wide aisles of the "Iron libraries," "the process *could* be repeated, putting in four more faces occupying 8 in. each, and still having the full American stack aisle of 32 in. left." He noted that "this increases capacity fivefold, and the electric light solves any difficulty from cutting off so much daylight." Finally, Dewey remarked that since his library had used "very comfortably" an aisle width as small as 26 inches, his suggestion did not seem to promote overcrowding.

In Snead & Company's 1915 bookstacks and shelving handbook, it was acknowledged that there were limits to how narrow aisles could be, however, for "although a person can pass through an aisle 21 inches wide, it is difficult to use the lower rows of shelves with aisles under 27 inches" because of the difficulty in stooping down in such a space. But, as mentioned earlier, even in stacks with the narrowest of practical aisles, they took up 65 percent of the floor space, leaving only 35 percent available for books. Rider's goal was to reverse that ratio, if possible, and in doing so he looked into another aspect of the history of compact book storage.

Rolling shelves were only one way to add moveable shelf space. Another solution was to install hinged bookcases. This is a variation on the doors to private library rooms that are fitted with shelves to have books completely covering all the room's walls. Such moveable shelves are sometimes real but more often are trompe l'oeil wallpaper or painting, as in the galleries of the British Museum Reading Room. When located in stacks, however, hinged bookcases are designed to swing out to reveal not a doorway but the permanently fixed shelves to which they are attached. Hinged cases could work well in stacks where the aisles are wide, so that the newer shelves could be swung fully perpendicular to the shelves to which one needed to gain access. In stacks where the aisles were exceedingly narrow, the hinged shelves had to be correspondingly narrow if they were to work. This concern would be exacerbated if a library wished to adopt the double-sided hinged shelves that Rider suggested to double the new shelving gained.

Sixty years before, in 1887, Dewey had gone Rider one better, anticipating the compact shelving of the later twentieth century. Speaking of suspended shelving units, he speculated, using the analogy of the card catalog, with which librarians would be familiar,

> Theorizers must not forget that this plan *could* be extended to infinity and to absurdity; e.g., a series of 100 double faces could be swung together, with one 75 cm aisle for the series. On an average it would be necessary to move 25 cases to transfer the aisle to the face needed, for a room full of faces corresponds to a drawer full of cards. One opens the cards at any point and makes room to read clear to the bottom by pushing the cards together on each side of the opening; i.e. the drawer has the space of one opening in its whole length; that space can be brought in front of any card desired, just as one aisle can be brought in front of any face.

Rider, who also used the librarian's term "face" to designate one side of a series of connected bookshelf sections, and who may be said to have been obsessive at times about finding an extra cubic inch in his stacks at Wesleyan, seems not to have picked up on Dewey's idea. Rather, Rider showed how existing shelf space could be made to hold almost twice as many books simply by rearranging them according to size rather than in strict order of classification. The trick, according to Rider, is to recognize where space is being wasted. It is not uncommon, for example, to find a shelf's height determined by only one or two tall books among a long line of shorter ones. Rider coveted the open space above these books, but he did not suggest that it be filled with books lying horizontally on the shorter volumes, as many a home librarian is wont to do. He was also very much aware that shelving books on top of books would make them susceptible to falling down into the space between shelves left for ventilation of the stacks. Of books so fallen, he observed that they "may be lost for years. Because our own stack decks also are not solid, books with us drop down through several decks—with disastrous effects upon their bindings!"

To avoid such disasters, and to save some of the space taken up by the thickness of shelves, Rider suggested that the steel shelves with which stacks were fitted be manufactured with minimum thickness. Manufacturers had long been motivated to do so, of course, to make their shelves economically attractive. But too-thin sheet shelves would not be very stiff, and so Rider suggested that a lip or "apron" be fitted on the back to stiffen the shelf. Such aprons were commonly formed around steel shelves by bending the steel sheet downward in both front and back. Rider suggested that the back apron might be formed by turning up the steel sheet, thus providing a "back stop" so that wider books could not be pushed onto the shelf behind, thus driving an opposite book out of alignment. The idea, like virtually all ideas, was not entirely new. In unsigned editorial notes on library shelving in an 1887 number of his journal, *Library Notes*, Melvil Dewey had elaborated on the "waste of space by giving too great depth to shelving." He described a law library (complete with his idiosyncratic phonetic spelling that would save space in books and journals as he wished to do in libraries):

> The back of each shelf is filled with a strip which leaves just space enuf for the standard law octavo. Then the shelves keep themselves flusht. In a miscellaneous library the back strip for padding must leave the standard space in front in order to accommodate all

> kinds of books. But a 7-inch space could be left regularly, and, if short blocks insted of one piece the length of the shelf ar used, when here and there a wider book occurred, a narrower block could be put in, thus giving needed room. Such a bibliothecal curb-stone helps wonderfully in keeping the library presentable where readers hav access to the shelves or where careless pages replace books on shelves needlessly deep.

In the tradition of Dewey, Rider suggested "a still more radical change in shelf design," and that was that the front of steel shelves be made with an upturned lip, and his observations here demonstrate the degree of detail encompassed by his thinking about the shelving of books:

> It would seem to be at least possible that we are at present making all our stack shelves upside down! Because steel shelves followed wooden shelves we have always taken it for granted that the bend of "stiffening" provided for them along their front edges has to be a down-hanging one, whereas it could just as easily be made a turned-up one. And it is possible that such a turned-up fold or "lip," on the upper side of the shelf, particularly if it were . . . a mere beaded roll, would provide us with a shelf much more effective than our present one. A turned up edge would tend to prevent our books from sticking out into the aisle, i.e. would automatically help keep them in alignment.

Rider recognized that, while an upward lip on the back of a shelf might work to keep the spines of uniformly wide law books aligned also along the front of the shelf, this would not be the case in a general library. Rather, in the latter the widely varying widths of books required a front lip over which they would have to be lifted but against which they could be pulled for the more or less automatic adjustment of their spines. But nothing is simple in bookshelf or any other design. Rider did acknowledge that a stiffening/aligning lip above a shelf's front edge might make it harder to take a book off the shelf, but he believed that objection to be canceled by the fact that it would be "slightly more easy to put it back there!" This may be taken as one of several examples of Rider's overenthusiasm for his redesigns of book storage details. Like many a myopic designer, he saw the advantageous aspects of his proposed changes to be much more positive than he saw their disadvantages to be negative features.

Having the spines right at the front edge of the shelf eliminates a dust-collecting surface before the books, of course, but it is merely relocated out of sight and perhaps out of mind behind them. I have come to believe that where relative to the front of a bookshelf the spines should stand is a matter of taste. I have continued to keep mine back from the edge for the most part, but I have also begun to experiment with pulling them all the way forward on some of the shelves to see what the fuss is about. The more I have lived with this latter arrangement, however, the more I have come to appreciate its appeal to its adherents. With less of the shelf visible, the books and not the shelves are more the focus of attention. On the other hand, it could be argued that the unseen fore-edges, and not the spines, should be aligned so as to provide as much lateral support as possible where the book's binding least provides it. This would, of course, mean a ragged line of spines, which no one seems ever to have advocated.

Among Rider's main concerns in the library space race was not where the edges of books belonged but how shelf area was wasted because the vast majority of books on a shelf were seldom as wide, i.e., deep, as the shelf itself. To use as much of this wasted space as possible, Rider arranged Wesleyan's books according to width and shelved them on their long sides, i.e., with their fore-edges resting directly on the shelf. Such a practice would have offended some nineteenth-century book lovers, who were advised, "Do not stand a book long on the fore-edge, or the beautiful level on the front may sink in." But in Wesleyan's library Rider seems not to have worried about such matters. The demands for space often require a book to be shelved with its long dimension horizontal, and in such cases whether it is better to do that spine up or spine down must be addressed. If spine down, the weight of the book rests on the spine, thus tending to compress it into a flattened shape. Shelving books spine up, on the other hand, allows the weight of the pages to pull down on the binding, causing the spine to "sink in." Either way damage can be done, but this is minimized if the book is held snugly between its neighbors, so that friction forces associated with the compression support the weight of the paper, at least in part, and keep it from either pushing or pulling excessively on the binding.

In any case, with books shelved horizontally according to Rider's scheme, where a book press section might have been seven shelves high, it now could be twelve shelves high, presenting an almost solid mass of books—or book bottoms. Through before-and-after photos of shelves so rearranged, the frontispiece of Rider's 1949 monograph shows 396 volumes of a long set of government documents that originally took up 3⅓ sections of shelf space rearranged into what he describes as 1¾ sections. In a general collection, composed not of government documents, which are by and large uniform in format, the books would have to be segregated according to size, something Rider and other librarians have also long advocated, to achieve a similar savings in shelf space.

Identifying books in the fore-edge down position between close-fitting shelves presented a similar challenge to that faced when medieval libraries shelved books fore-edge out, but Rider expected to solve the problem by writing each book's title, author, call number, and other relevant information on the bottom of the book, which was the part of the book facing the aisle, or as Rider put it, "presented" to the patron or staff member looking for a particular item. Rider soon found that many a book's bottom did not present a smooth enough surface on which to write easily, however, and so he resorted to using a guillotine paper cutter to trim to a smooth surface those books that needed it. Still, some books eluded the pen, such as those that were too thin. Thus he resorted to boxing many of the books, which then presented a wide enough flat-bottom surface so that a typed label could be affixed. Though Rider might have been prone to exaggerate the space savings of his methods, as when he minimized the space that his pamphlet-containing boxes took up on the shelves, overall his analyses were sound and truly space-saving, even if a bit extreme and labor-intensive.

Rider acknowledged that he was frequently asked whether increasing the storage capacity of his stacks by as much as 60 percent would not overtax the structure built for more conventional book storage and thus dangerously overload the stacks. His response to "this very proper query" was that engineers had designed the stacks with a structural "margin of safety" that was as high as 300 or 400 percent, i.e., a factor of safety of 3 or 4, in the case of stacks manufactured by the Snead company, which in practice actually made them even stronger than that. There is a factor of safety in all library structures, of course, though it may not be as high as the 4 that was required by the University of Minnesota Library contract under which metal bookstacks were bought in the 1920s.

It certainly is the case that all properly engineered structures are designed and built to be stronger than their rated capacity, but what Rider did not point out was that by compacting his books he was reducing the margin of safety to the extent that he increased the book load on the structure. Though calling on the engineering reserve may seem a clever way to get more out of a structure's strength and space, it is ill advised to reduce the factor of safety that was integral to the original structural design, to account for variations in material strength, misalignment of supports, poor workmanship, and other contingencies of construction, maintenance, and use. In Rider's situation it might not have jeopardized the structure's integrity, but that might not always be the result.

It is unlikely that any rearrangement of books on existing shelves can save more space than Rider accomplished, but a different approach to the problem was suggested as early as 1890—not by a librarian, but by a past and future prime minister of Britain, who wished to "stuff into a single room enough books to fill another man's house." William Ewart Gladstone believed that "in a room well filled with [books], no one has felt or can feel solitary." He also held that the purchase of a book did not end with "the payment of the bookseller's bill," for such payment was "but the first term in a series of goodly length," which involved building and maintaining bookcases, dusting, and cataloging: "What a vista of toil, yet not unhappy toil!"

Gladstone had strong opinions about how to shelve books, and he disposed "with a passing anathema of all such as would endeavour to solve their problem, or at any rate compromise their difficulties, by setting one row of books in front of another." No Pepysian bookcases would do for the prime minister, and so he had to determine an alternate scheme to accommodate his books, which he felt should address three criteria: "economy, good arrangement, and accessibility with the smallest possible expenditure of time." He believed that books should be "assorted and distributed according to subject," but he had to admit that the criteria were not independent of each other, for "distribution by subjects ought in some degree to be controlled by sizes. If everything on a given subject, from folio down to 32mo, is to be brought locally together, there will be an immense waste of space in the attempt to lodge objects of such different sizes in one and the same bookcase."

Gladstone did not approve of highly ornamental bookcases, which were in fashion in Victorian times: "Now books want for and in themselves no ornament at all. They are themselves the ornament." He advocated the stall system, "after the manner of a stable," which had by and large given way to the more genteel wall system in private libraries. Being a stickler for detail, Gladstone argued for the fixed shelf, for "it contributes as a fastening to hold the parts of the bookcase together," and he even specified based on his "very long experience" the size of lumber to be used: 3-foot-long and 12-inch-deep "shelves of from half to three-quarters of an inch worked fast into uprights of from three-quarters to a full inch will amply suffice for all sizes of books except large and heavy folios."

By way of a footnote added "in illustration" to his article on books, Gladstone outlined how "nearly two-thirds, or say three-fifths, of the whole cubic contents of a properly constructed apartment may be made a nearly solid mass of books," by employing what we would describe as compact shelving:

> Let us suppose a room 28 feet by 10, and a little over 9 feet high. Divide this longitudinally for a passage 4 feet wide. Let the passage project 12 to 18 inches at each end beyond the line of the wall. Let the passage ends be entirely given to either window or glass door. Twenty-four pairs of trams run across the room. On them are placed 56 bookcases, divided by the passage, reaching to the ceiling, each 3 feet broad, 12 inches deep, and separated from its neighbors by an interval of 2 inches, and set on small wheels, pulleys, or rollers, to work along the trams. Strong handles on the inner side of each bookcase to draw it out into the passage. Each of these bookcases would hold 500 octavos; and a room of 28 feet by 10 would receive 25,000 volumes. A room of 40 feet by 20 (no great size) would receive 60,000. It would, of course, be not properly a room, but a warehouse.

Essentially the same idea was proposed in 1893 by the librarian of the university library at Glasgow, who, in response to the recently described sliding presses at the British Museum, asked,

> Why should not all the presses in a library be brought into close proximity, side by side with each other, so that each one would require to be pulled out into an open space when access to its shelves is wanted? Of course in this arrangement the movement would be endwise, and not, as in the British Museum, frontwise.

Such an arrangement, acknowledged to have been inspired by installations at the Bodleian, British Museum, and Glasgow libraries, was developed with the help of a "versatile engineer" from a stack construction company and installed in the Toronto Central Circulation Library around 1930. From the beginning such a system, which increased book storage capacity by over 40 percent, was seen to provide as a bonus "considerable freedom from dust" on the books and "from the exclusion of light the bindings would remain longer fresh and the edges of the leaves would perhaps escape the brown tinge which so frequently discolors not a few volumes." But the scheme of sliding or rolling shelves lengthwise into a wide aisle did not catch on, partly because of the problem of books sticking out of a shelf catching on others in a facing shelf.

The idea of another kind of moving shelf did gain currency later in the twentieth century. Modern compact shelving that moves transversely is almost always of the rolling or sliding kind, and, unlike the British Museum shelving that rolled along tracks hung from the ceiling, is supported from tracks or rails located beneath the shelving. The essential idea behind compact shelving is to reduce the wasted aisle space to almost nothing. In fact, the system is often called moveable aisle compact shelving, with only an aisle provided between the shelves that are being consulted, much as Dewey described in his analogy with the card catalog. The shelving units in their compacted form stand close together with barely enough space to squeeze a hand between, let alone get a book out. The units are often powered or mechanically advantaged in some way or another through gearing, and thus they are easy to move away from each other when access is desired along the aisle that is opened up for the purpose. For safety reasons and so as not to dislodge books by sudden starts and stops, the shelf sections do move slowly, however, and elaborate safeguards, including electronic floor sensors, necessarily have to be provided so that the shelving units do not close upon and crush a patron or library staff member.

Rolling book presses were installed in the Toronto Central Circulation Library around 1930. They rode on "heavy rubber composition wheels" and were "ridiculously easy to move" longitudinally out into the stack aisle when access to their shelves was needed.

Compact shelving units, which often hold less-used materials for which there is not a high demand, can be filled end to end, top to bottom, with books, thus utilizing virtually 100 percent of their available shelf space. Regular library bookstacks with traditional fixed aisles are generally considered loaded to capacity when physically they might be less than 90 percent full. This is because of the necessity to add newly acquired books, that by traditional common cataloging practice must be inserted between books already shelved. Since the new books may have to be located at any place in the collection, there everywhere has to be empty space on the shelves to accommodate the books lest the whole collection require constant reshelving on a massive scale.

Under normal conditions, when library shelves are only about 60 percent filled, they will allow for a goodly period of accumulation

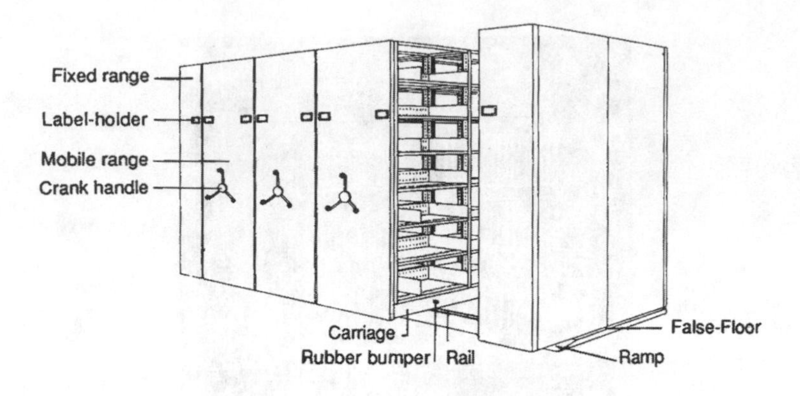

Modern compact shelving moves perpendicular to the direction of the shelves and opens up one aisle at a time.

before a major rearrangement of the books in the stacks is necessary, but it is recommended that planning be undertaken at that time for adding shelving. According to accepted library practice, the point at which it will become difficult to interpolate new books into the existing shelf arrangement is when the shelves are 84 percent filled. It is at that point that the entire collection must be opened up like an accordion to begin the inevitable closing in anew. There will be room to effect the expansion into new shelf space only if planning for additional shelving had begun in time. If such planning does not proceed to the realization of new space, books often have to be stored on the floor before bookcases or, where cases have tops, above the cornice.

When iron and steel bookstacks became passé and library stack floors started again to be built to support bookcases rather than being supported by them, the ability to install truly compact shelving that covered every square inch of floor space (save an aisle or two for access) became the ideal. However, the additional load that would have amounted to very nearly 200 percent of the original might have been too much for a light structure to take. Library buildings do have to conform to local building codes, which require a certain margin of safety, but it cannot necessarily be assumed that the engineer designed into the structure any more reserve strength than that. To install compact shelving on an existing floor already being used to its design capacity is to overload it illegally and unwisely. Hence, the installation of compact shelving in existing library buildings may be limited to basement or subbasement floors that were extra strong to begin with and so can accommodate the added weight.

Where there is no more room or floor capacity in conventional stacks to erect new or compact shelving, or no more resources or desire to expand the capacity of a library building, off-site storage is often resorted to. In this case, a building never intended for housing a library is typically used to hold books in a warehouse-like configuration, with tall sections of shelves, which require ladders or other climbing aids to reach. The idea of a separate storage site for less frequently used books was promoted in the late nineteenth century by Charles William Eliot, who was president of Harvard College from 1869 to 1909, and who held that "a five-foot shelf would hold books enough to give in the course of years a good substitute for a liberal education in youth." (He later wrote an introduction to the Harvard Classics, which came to be known collectively as "Dr. Eliot's five-foot shelf.") President Eliot had "called attention to the urgent need for an addition to Gore Hall, which had served as the library building since 1840 and was overcrowded," and the library was added to in the late 1870s and again in the late 1890s, as we have seen, which bought time until a new building could be funded and built. In the meantime, books had begun to be stored in the basements of other buildings.

By 1885, Eliot was pointing out that not all the books in Harvard's library were equally in demand. In fact, in a given year there appeared to be only a small proportion of books that were actually called upon and used. It was around the turn of the century that he set out the problem as he saw it:

> One who watches the rapid accumulation of books in any large library must long for some means of dividing the books that are used from those that are not used, and for a more compact mode than the iron stacks supply of storing the books that are not used. Although the iron stack was a great improvement on any former method of shelving books in a large library, it still wastes much room, and access to the books that are wanted is made slower and more difficult by the presence on the shelves of a great number of books that are never wanted. The devising of these desirable means of discrimination and of compact storage seems to be the next problem before libraries.

Eliot's observations would seem to some to have been incontrovertible, and he put them before scholars and the library profession in the form of a proposal for "depositories of dead books," in which volumes "not in use" would be stored. (A century earlier, a Spanish priest experiencing great difficulties in gaining access to the treasures of the Vatican Library called it "a cemetery of books." Gladstone employed the term "book-cemeteries" for compact shelving programs. Neither term endeared such schemes to book lovers.) "The modern steel stack," Eliot wrote, "is not a decorative or inspiring structure, and we should all be glad to advocate with good conscience more beautiful and interesting forms of construction for the library of books in use." For those books not in use—the dead ones—he suggested that less beautiful or interesting depositories might be located in Washington, New York, and Chicago. These deposit libraries would be expected to make books available widely, of course. He answered the objection that such an arrangement would slow down the delivery of books:

> The student and the general reader alike should be willing to await the delivery of the book he wants for hours and even days, just as a naturalist waits for the season at which his particular material is to be found, or for the time of year when his plant flowers, or his moths escape from the chrysalis, or his chickens or his trout hatch. The real student ought to be capable of some forelooking, and of a certain deliberation in reading.

Eliot considered books "sufficiently accessible if they could be delivered within twenty-four hours," and opposed spending millions of dollars on storage facilities just so readers could have their requests filled in minutes. He considered browsing an afterhours pastime, not an essential means of scholarship, for he argued that consulting books in library stacks was an "unscientific" approach because no collection was in fact complete. He proposed that all libraries in the Boston area, for example, should store their disused books in a common warehouse, with duplicates discarded. He was opposed to classified storage in such a facility, because it was wasteful of space, and he advocated the shelving of books by size. His philosophy can be summarized in an oft-quoted passage from his 1900–1901 annual report:

> It may be doubted whether it be wise for a University to undertake to store books by the millions when only a small proportion of the material stored can be in active use. Now that travel and the sending of books to all parts of the country have become safe and cheap it may well be that great accumulations of printing matter will be held accessible at only three or four points in the country, the great majority of libraries contenting themselves with keeping on hand the books that are in contemporary use, giving a very liberal construction to the term "contemporary." . . . An examination of the book(s) once in five or ten years might divide the unused from the used. The unused might be stored in a much more compact manner than they are now, even in the best arranged stacks. . . . Such a division of the books in a library is repulsive to librarians and to many learned men who like to think that all the books on their respective subjects, good, bad, and indifferent, alive and dead, are assembled in one place. In a university, however, the main object of a library must always be to teach the rising generation of scholars. Whatever injures a library for the use of learning's new recruits should be avoided, but without making it impossible for the library to serve also the needs of veteran scholars.

Some scholars were indeed aghast at the suggestion of off-site storage, and they argued that they used a lot more books than surveys acknowledged. One Harvard professor "enforced the point by writing the date and his initials on the volumes he consulted, as a warning against sending them away." The idea of off-site or warehouse storage of books was supported by few librarians, and it was not pursued at the time. The principal development that made the topic a dead issue, at least at Harvard, was the gift that enabled Gore Hall to be replaced with the Widener Library, completed in 1915. However, it was not long before shelf space reappeared as a problem, and it became inevitable that something like Eliot's scheme had to be implemented. It was during the 1930s and 1940s that deposit libraries were increasingly discussed as necessary options. The New England Deposit Library, in which Harvard participated, opened in 1942.

At about the same time the New York Public Library also began to store some of its books off-site. Within years, the idea caught on, and by the late 1940s even smaller libraries were participating in cooperative efforts to keep some of their lesser-used volumes in a separate location. One such effort was the Midwest Storage Warehouse in the Chicago area.

In time some large research libraries set up their own book warehouses. When I last visited it, the Duke off-site storage facility, located about a mile from the main library in a metal building among other metal buildings near the railroad tracks, had heavy industrial shelving reaching from the concrete floor almost to the high corrugated-steel roof. The books were crowded into the shelves like the pre-Christmas inventory in a toy store, and books were stacked behind books and on their fore-edges as on Rider's compact shelves at Wesleyan. Those who retrieve books in such an environment are clearly adept at moving among the giant bookcases, but it is not a situation that is reader- or browser-friendly.

Taking matters one step further are the automated storage and retrieval systems that utilize the late-twentieth-century development of computer control. In this arrangement, shelving no longer resembles that of bookcases but is of the industrial-rack kind encountered in food warehouses and hardware superstores. The shelving need not even be solid, for the books are stored in bins that rest on shelves as high as 40 feet off the floor. The books in each bin are kept track of through a computer, which also directs a forklift-like retrieval device that moves on rails along the 90-foot-long aisles between the shelving. When a book is requested, the bin in which it resides is brought to the operator or attendant, who then removes the appropriate title from the bin and returns it to its place at the push of a button. There are few theoretical limits to size for such extreme efforts to fit as many books as closely together in as little floor space as possible, but as warehouses grow larger so does their rent and cost of operation, and so librarians have long been attentive to alternatives even to the compact storage of physical books.

The technique of microfilming books developed between the First and Second World Wars, and for a time it promised to be the solution to the problem of ever-growing bookstacks and book storage facilities. The introduction of microfilm generally was expected to be as revolutionary as printing from moveable type, with the new form driving out the old. Readers and libraries, it was predicted, would in time use microforms instead of paper books, with reading material projected upon walls so that groups of people could enjoy books the way they do motion pictures. Such projected books were, in fact, available for hospitalized veterans in the wake of the Second World War, but "experience began soon to show that there was great resistance from readers to the inconvenience of having to read through an apparatus."

In the last decades of the twentieth century, the computer virtually replaced, or at least halted the growth of, the card catalog in libraries large and small, as Nicholson Baker has chronicled. The threat of the computer to displace reference books was incompletely fulfilled by the end of the century, however, and the computer has presented new space problems. In my own university library, computer terminals occupy growing amounts of floor space, but the hard-copy reference collection still remains a vital part of the mix. Some older patrons, at least, still prefer not to have to use any apparatus other than a conventional book. Some wags have called the CD-ROM the "new papyrus," and see it as marking a new era in the history of mankind, one called CD in allusion to BC and AD.

What the future actually will hold with regard to computerized databases replacing traditional books is still anyone's guess. The next several decades will reveal whether a generation that grew up with computers will eschew the conventional book entirely for the electronic version. If this does happen, some libraries at least are likely to scan and digitize their collections and order all new books in compact-disk or other electronic format. Such a scenario would mean plenty of available shelf space as old books in conventional form are discarded and few new ones acquired. And the shelving of uniform-sized compact disks would be a dream fulfilled to librarians like Fremont Rider.

Another scenario is also possible, and that is that the e-book will succeed and that books will be downloaded from the Internet. But at the same time, it may be the case that the digital network and the terminals that tap into it will become saturated as limits to growth of computer memory and speed of operation are reached at the same time that electronic traffic becomes gridlocked with e-mail and World Wide Web use. If that were to happen, there would likely be pressure to keep older books in print form, and perhaps even continue to issue newer books that way, rather than clutter the Internet with more and more information. Under such a scenario, older books might not be allowed to circulate because so few copies of each title will have survived the great CD digital dispersal, leaving printed editions that will be as rare as manuscript codices are today.

In spite of potential problems, the electronic book, which promises to be all books to all people, is seen by some visionaries as central to any scenario of the future. But what if some electromagnetic catastrophe or a mad computer hacker were to destroy the total electronic memory of central libraries? Curious old printed editions of dead books would have to be disinterred from book cemeteries and rescanned. But in scanning rare works into electronic form, surviving books might have to be used in a library's stacks, the entrance to which might have to be as closely guarded as that to Fort Knox. The continuing evolution of the bookshelf would have to involve the wiring of bookstacks for computer terminal use. Since volumes might be electronically chained to their section in the stacks, it is also likely that libraries would have to install desks on the front of all cases so that portable computers and portable scanners could be used to transcribe books within a telephone wire's or computer cable's reach of where they were permanently kept. The aisles in a bookstack would most likely have to be altered also to provide seating before the desks, and in time at least some of the infrastructure associated with the information superhighway might begin again to resemble that of a medieval library located in the tower of a monastery at the top of a narrow mountain road.

CHAPTER ELEVEN

The Care of Books

Richard de Bury wrote in the *Philobiblon* of "the delicate fragrant book-shelves" of Paris. Five centuries later the French were still admired in bibliological circles as "the teachers of Europe," as they were characterized in a Victorian self-help book that gave the "amateur" collector practical advice on how to care for books:

> There was once a bibliophile who said that a man could only love one book at a time, and the darling of the moment he used to carry about in a charming leather case. Others, men of few books, preserve them in long boxes with glass fronts, which may be removed from place to place as readily as the household gods of Laban. But the amateur who not only worships but reads books, needs larger receptacles; and in the open oak cases for modern authors, and for books with common modern papers and bindings, in the closed *armoire* for books of rarity and price, he will find, we think, the most useful mode of arranging his treasures. His shelves will decline in height from the lowest, where huge folios stand at ease, to the top ranges, while Elzevirs repose on a level with the eye. It is well that each upper shelf should have a leather fringe to keep the dust away.

The "long boxes with glass fronts" are of course the lawyer's and barrister's sectional bookcases that came into prominence during Victorian times and that would command such outrageous prices on the antique market a century later. (My wife and I were fortunate to find a fine set of these, in tiger-oak veneer and with claw feet, before they were so fashionable, and I have indeed moved them, section by section, from place to place with their books in them. Unfortunately, most of the sections were made for the smaller books popular in the nineteenth century and not the larger octavos that comprise so many general books of the late twentieth. Today, barrister's cases are again made, but in the modern mode, with straight lines, plate glass, and flat feet, and they are used for storing everything from paperback books to CDs and videotapes.)

Assuming books "of rarity and price" would tend to be older volumes, the Victorian prescription of a closed armoire to hold them was historically apt, of course. As for the Elzevirs, these would have been small volumes that would have gone "readily into the pocket" and thus easily "smuggled" into houses where not every family member appreciated the extravagance of spending money on books and filling bookcases with them. The Elzevirs were a Dutch publishing family dating from the late sixteenth century that came to be known for its publication of duodecimo editions pirated, especially from the French, in times "of slack copyright." The diminutive books that took the name of their publishers were "beautiful, but too small in type for modern eyes." As for "the leather fringe to keep the dust away," it also served to mask the ragged line that results when books of different sizes were shelved among the uniform Elzevirs. This skyline effect disturbs some book owners even today.

While dust on books and shelves may be an annoyance, bookshelves themselves can be the enemies of books, for problems of light, climate, and critters can be even more damaging. As everyone knows who has left shelves of books in a brightly sunlit room over the years, their spines and dust jackets can get badly faded. Where shelves hold books of different heights, as they often do, the top of a taller book next to a shorter one can fade to look as if it has acquired a two-tone binding reminiscent of the automobiles of earlier generations. Window shades can be drawn to prevent such things from happening, of course, but some people are torn between keeping their book bindings bright and keeping a room bright. In my own case, I find myself opening the window blinds wide in the winter to let maximum sunlight into an otherwise dreary room, but this also lets the low southern sun irradiate my books as it makes its low transit across the sky each day. One book owner solved the problem by installing window shades not on his windows but on his bookcases. Another "won't let his wife raise the blinds until sundown, lest the bindings fade." This same collector, who is an investment analyst by day, "buys at least two copies of his favorite books, so that only one need be subjected to the stress of having its pages turned."

More independently wealthy collectors, such as Paul Getty of Oxfordshire, England, do not risk their books being exposed to unchecked sunlight or even being touched by painted or stained wooden shelves. In Getty's "castlelike folly" of a library building, the skylights "are treated to screen out ultraviolet rays," and even the library's electric lights are on dimmers. The shelves are "lined in billiard-table baize, so a book is not marked when removed." Getty has also gone to extra lengths to protect his books by having holes "placed in the backs of the shelves to circulate cool air around the books while leaving the main body of the room warm enough for human comfort." The director of the collection—and a book collection in a castle, even or perhaps especially a folly of one, certainly warrants a director—has pointed out that central heating is no friend of books, and the cooler they are the better for them. Furthermore, "books, like wine, need to be kept at a regular, unfluctuating temperature." The Getty library is also fitted with a sprinkler system—in case of fire—but the sprinklers do not connect to water pipes as they do in many a public and research library but to a source of Halon gas that would deprive a fire of oxygen without getting the books wet. In more modest libraries one usually simply hopes that fires never start.

Bern Dibner, the electrical engineer, inventor, and premier twentieth-century collector of books in the history of science and technology, kept his treasures in wooden bookcases with glass doors in the offices of his Burndy Engineering Company. Since the Burndy factory, which manufactured electrical connectors, was fitted with a water-sprinkler system, the rare books were in danger of being soaked if the system was ever triggered. To protect his collection in this event, Dibner had the bookcases fitted with metal canopies to shed the water as a pitched roof does.

Getty's shelves and Dibner's precautions are extremes, of course. Domestic bookcases generally have not evolved as far as those in the Getty library, and certainly not as far as the stacks and compact storage systems of institutional libraries, for a collection of books in the home is often much more limited in its maintenance budget and in its numbers. Whereas a research institution's library derives its strength from what it retains—virtually everything but duplicates—the home library can continually concentrate its essence by the selective discarding of older books to make room for the new. The process is known as weeding or editing a collection, and it can be driven more by the shelf space one has than by any inclination toward perfection in one's holdings. Every home library does seem to have a certain core collection that is not dispensable, however.

There are exceptions, of course, and I have known younger collectors especially who seemed to think of themselves as budding Librarians of Congress. These individuals seem never to discard any book, but rather build more cases as their accumulations grow. And, as it is with many bibliophiles on a limited budget, the books seem to be much more important than the appearance of the bookcases. One acquaintance, a part-time farrier who carried his coals and irons in the back of a small pickup truck, had so many books that he had filled every single wall space in what must have once been a living room with shelves of the kind I would expect to find in a basement or garage. With the walls covered, he also located shelves in the middle of the room, so that one had to wend one's way through them as if in a garden maze. The house he and his wife lived in was modest, and they must have spent all their spare cash on books and the shelves on which to store them. Indeed, he appeared to have taken up horseshoeing to support his habit for books and the cases to hold them. I once asked him how he had come to practice his trade, and he informed me that he had read about it in a book. Ironically, however, the physical exertion of his job left him little time or energy to read his books or write the ones of his own that he wished to add to the shelves of libraries everywhere.

The tendency to let one's books take over one's living space, if not one's life, is not all that uncommon, as demonstrated in the delightful if quirky off-the-shelf coffee-table volume *At Home with Books,* which provides glimpses into the homes of book people from all walks of life. The New York City apartment of the poet and translator Richard Howard, for example, appears to resemble more a bookshop than a home. According to Howard, he "really wanted to be a reader, not a writer," and his floor-to-ceiling, door-to-door bookshelves filled to overflowing cannot lead anyone to doubt the assertion. Roger Rosenblatt, another writer living in New York, who once performed a one-man show entitled *Bibliomania,* has "made room for books in virtually every room of the house," including the dining room.

Interestingly, unlike Rosenblatt's bookshelves, which appear to be a hefty 1½ inches thick, almost a little too heavy-looking for the delicate dining-room chairs they surround, Howard's long slender shelves appear to be no more than 1 inch thick, if that, and seem to be sagging here and there under the weight of their burden. Rosenblatt's shelves may indeed be thicker than they have to be to hold a straight line, but they will not sag, and they would probably not appreciably sag even if they were longer than they are. Howard's shelves, on the other hand, are clearly not up to the task. They appear to be awfully long, and they might indeed sag, if not slip off their ends, if it were not for the support of the books stuffed onto the shelves below. The problem of reducing a shelf's sag to acceptable proportions is arguably a matter of taste, but too much sag is definitely unsightly and can make the shelves unsettlingly insecure-looking. Usually, however, we barely notice the physical dimensions of shelves because we focus on the books or how they are used.

I once attended a dinner party in one of the rare downtown Houston high-rise apartment buildings that just as easily could have been built in New York or another large city. The living and dining areas of the place comprised one large open area in the corner of the building, looking out over a park and the low structures that surround it. A windowless space on one outside wall was covered by a single floor-to-ceiling, window-to-window arrangement of bookshelves that were, naturally, filled with books. The dining table sat directly in front of this arrangement, which supported distinctive and familiar period spines of the likes of the yellow-and-black Scribner's paperback edition of *The Great Gatsby* and the wine-colored Vintage edition of *Finnegans Wake,* thus at the same time dating their owners as having been students in the 1960s and strongly suggesting that one of them at least might well have been an English major in college. The shelves had an especially high clearance from one to the other because these patrons of the arts also had a good number of coffee-table-size art books standing on their shelves.

Dinner was served buffet style, and we ate at the formal dining-room table set with beautiful silver and handsome crystal. Besides the incongruity of eating fried chicken—Texas style—among such elegant trappings, what must have surprised the guests the most was the makeshift place mats on which we were expected to set our plates. Instead of the usual rectangles of cloth or textured weaves of grass there were art books, each opened to a two-page spread of color and composition. My assigned seat had me eating over a Monet, one of the sprawling water lily canvases. The books appeared to be the kind one gets from a Time-Life book club, which book snobs might consider all but disposable, like paper place mats, but the very thought of using books of any kind as table protectors made not a few diners uncomfortable. All wishing to be polite company, however, none made a scene or, after a respectable double take, refused to set their plates down. Whether the greasy crumbs were shaken out and the books reshelved after the dinner party, I do not know. Perhaps any soiled pages were disposed of, the way Sir Humphry Davy tore out the pages of his books as he read them.

Most shelves are not in the dining room, of course, and the books on them are not to be touched with dirty fingers, much less eaten over. We do not have to read it in the *Philobiblon* to know of books that "great injury is done so often as they are touched with unclean hands." The telltale trails left by unhygienic users were more graphically described by a member of the staff of the British Museum Reading Room, who recalled on his first day "being shown by the horrified Superintendent of the Room the coffee-coloured mark traced across a page of print by the forefinger of a querying reader." But the same staff member may have reveled in the breakfast served off the catalog desks the day the Reading Room opened. It is not only by hands and food that books can be soiled, however, and de Bury believed that "the race of scholars is commonly badly brought up":

> You may happen to see some headstrong youth lazily lounging over his studies, and when the winter's frost is sharp, his nose running from the nipping cold drips down, nor does he think of wiping it with his pocket-handkerchief until he has bedewed the book before him with the ugly moisture. Would that he had before him no book, but a cobbler's apron! His nails are stuffed with fetid filth as black as jet, with which he marks any passage that pleases him. . . .
>
> He does not fear to eat fruit or cheese over an open book, or carelessly to carry a cup to and from his mouth; and because he has no wallet at hand he drops into books the fragments that are left. Continually chattering, he is never weary of disputing with his companions, and while he alleges a crowd of senseless argu-

> ments, he wets the book lying half open in his lap with sputtering showers. Aye, and then hastily folding his arms he leans forward on the book, and by a brief spell of study invites a prolonged nap; and then, by way of mending the wrinkles, he folds back the margin of the leaves, to the no small injury of the book.

The narrator Adso in Umberto Eco's medieval mystery *The Name of the Rose* was similarly offended by how the use of books injured them. He compared books to "a very handsome dress, which is worn out through use and ostentation":

> Its pages crumble, its ink and gold turn dull, if too many hands touch it. I saw Pacificus of Tivoli, leafing through an ancient volume whose pages had become stuck together because of the humidity. He moistened his thumb and forefinger with his tongue to leaf through his book, and at every touch of his saliva those pages lost vigor; opening them meant folding them, exposing them to the harsh action of air and dust, which would erode the subtle wrinkles of the parchment, and would produce mildew where the saliva had softened but also weakened the corner of the page.

Whether or not they dog-ear the corners of pages, moisten their fingertips, or use their napkins properly, it may give the wrong signal to guests to have bookshelves in the vicinity of arguments or food. I have long been confused by the mixed signals some institutions give to patrons about eating in the library. Though signs may state clearly that no food or drink is to be brought into the building, there seems seldom to be general compliance or any strict policing of what in fact can be brought through the entrance. Perhaps it was the installation of airport metal detector–like electronic gates that beep and lock when someone tries to leave with an uncleared book in a backpack that has given library staffs a hands-off attitude. For whatever reason, however, food and drink of all kinds often do end up in the stacks and carrels of a library, and its aisles can smell more like the alley behind a restaurant than a corridor among bookshelves.

The practice of lining with plastic bags the wastebaskets that are placed at every open and in every closed carrel in a library seems also to encourage the disposal of banana peels, half-empty drink cans, and all sorts of more exotic foods. With the addition of wastebasket liners—added perhaps because not everyone felt that wax and brown paper should separate mustard-laden lunch meat from a steel wastebasket, or perhaps because the cleaning crew found it more convenient to pick up a bag of trash out of a wastebasket and take it to the hallway where the cleaning cart was than to carry the wastebasket out, empty it, and then bring it back into the office—things changed. For whatever reason they were introduced, plastic bags in wastebaskets seem to encourage sloppy disposal habits. Given the signals present in the stacks, is it any wonder that students do not take seriously the NO FOOD OR DRINK signs at the library entrance?

I once worked in a carrel that was in a wonderfully quiet corner of the library. The only sounds were the background noise of the air-conditioning system and the occasional carrel door being locked or unlocked and a chair being pulled up to or pushed away from a desk. Most of the carrel users were exceedingly quiet otherwise, but around lunchtime there began increasingly to be the rustle of lunchbags and, in time, the popping open of what sounded to me like Tupperware. The odors that emanated from those carrels were foreign to my lunchtime nostrils, but they reminded me of the over-marinated salads that a one-time officemate used to unseal every noon over the *Wall Street Journal*, which he used as a placemat of sorts. It apparently made little difference to him if he spilled dressing on the paper, because he had read it on and off so much throughout the morning that I could not imagine that there was anything in it that he had not already digested.

Library shelves, being themselves generally unequipped with waste baskets, lined or otherwise, provide no evident place to dispose of waste. Apparently being too considerate to throw trash on the floor, however, library patrons seem all too often to leave candy and gum wrappers, and more, on the shelves themselves, and sometimes in the books as bookmarks. Such behavior would no doubt have offended but probably not surprised Richard de Bury five centuries ago. The more things change, the more they remain insane.

There are, of course, other things than food to worry about in presenting our bookcases, which themselves often can be read like a book. Squeezing books too tightly together on a shelf may seem to book lovers akin to squeezing too many people into a crowded bus, subway, or elevator. There is an incivility about it that, had it been done in his time, Richard de Bury might well have written of as he did of the young reader with dirty fingernails and runny nose. A related matter of shelving etiquette has to do with squeezing onto a shelf books that are a hair too tall for it. And purists might even add that storing books horizontally across the tops of a shelf-full of vertically ordered books in a Stonehenge-like arrangement is something that simply should not be done. It is, however, tempting to put to good use the space that otherwise might go to waste. To minimize damage, some will choose carefully the location atop a shelf of books to place another horizontally, preferring a flat stretch above volumes of uniform height so as to distribute the weight of the interloper across as many hardbacks as possible. No matter how carefully or casually done, however, the practice of laying books horizontally across the tops of vertical books also buys time before the owner has to thin out the collection or rearrange the shelves—often only after adding more.

The accumulation of books on shelves appears to be inevitable, and the search for ever more places to store books appears to be without limit. The house or apartment with too many books seems always to acquire even more. Shortly after my daughter had moved into a new apartment, there was room on her shelves for her kitten to jump up on them and find a comfortable place to sleep. That is no longer the case, not only because the kitten has grown into a sizable cat but also because the collection grew to fill all possible voids in the bookcases.

In time, the tops of freestanding cases begin to collect books instead of dust. And as more and more shelves are added, as they often are, rooms, hallways, and stairways begin to narrow. According to the widow of one New York City collector, their eighteen-room apartment "was so crowded with books that her stepchildren had to walk sideways down the hall to get to their (book-lined) bedrooms." When hallways and bedrooms are exhausted, the space beneath tables may begin to fill in with books, the table legs sometimes serving as bookends. Some accumulators, who often call themselves collectors, have been known to pile books in the middle of a room and place a board or piece of glass atop them and call it a table—a book-coffee table on which to place coffee-table books.

No matter how grand or common they are, whenever houses and apartments are vacated, the books are taken off the shelves and sent to, one usually hopes, better shelves. The empty shelves left behind in the vacant space give an eerie feeling to many with books, for so much unoccupied shelf space seems to be an unnatural phenomenon. Indeed, if nature abhors a vacuum, most book lovers seem to abhor an empty shelf, or even a narrow gap in one, judging from their propensity to keep buying new titles. One book accumulator's wife puts a positive spin on the situation, for she sees an empty shelf as a welcome thing because it provides room for more books.

Some owners of books—especially of those investment properties known as "rare books"—apparently believe that volumes on a shelf are like paintings on a museum wall, there to be seen but not touched. One book collector was reported to have shouted from across the room, "What are you doing?" when a visiting friend of one of his children began to remove a volume from the shelf. The sentiment is not new, for the nineteenth-century English essayist Charles Lamb called book borrowers "mutilators of collections, spoilers of the symmetry of shelves, and creators of odd volumes." Some book owners see "every hole in the shelf a crater."

When the craters have been filled, kitchen and pantry cabinets can be commandeered in the fight to find bookshelf space, and a family's eating habits can be changed. When the china is displaced by paper plates, there is no longer any reason why books cannot be stored in the dishwasher too. An empty refrigerator is an excellent repository for the most valuable of books, because books like low temperatures best. As long as the power does not fail, no mold or mildew can grow, and no insects can breed. With the kitchen full of books, the closets can be eyed. But clothes are less easily done without than fresh food. (Book people seem to love to dine out and talk about books.) Storage space can always be found in the fullest clothes closet, however, by giving away things that have not been worn in the last week or so and by compressing the rest. In short, even the most crowded homes and apartments always have room for more books, though that space might not be in the form of traditional bookshelves.

In spite of the cleverness and adaptability of librarians and book accumulators to find nooks and crannies for storage, conventional bookshelves are still by far the preferred way to hold and display books. Yet for all their simplicity and directness of purpose, there are some practical construction principles that must be followed if the shelves are to function as we wish. They must be deep enough, for example, lest our largest books hang over the edge, like long pieces of lumber hanging out the back of a pickup truck, requiring a bright red or yellow flag to warn us of their projection. In a bookcase, the shelves must also be amply spaced vertically, lest our tallest books not fit. Such considerations have made the better bookcases more or less uniformly deep and fitted with adjustable shelves. Sometimes, however, because we want to economize or express our individuality, we need or want bookcases that are different from the run of the mill.

Libraries, whether in the private home or apartment or in an institution, always present a dilemma whenever bookcases along two perpendicular walls meet in an inside corner. There are several options, of course, including leaving hidden and unused the space behind where two intersecting bookcases clash. Sometimes, especially with freestanding cases that cannot be cut to fit, the end of one is hidden behind the end of the other. If the bookcases do not fill the wall space anyway, the end of the second case can be positioned a book's width short of the other, which itself is carried to the wall. This leaves room to reach in and shelve and retrieve books from the dark cul-de-sac, but this is generally an unattractive solution. Library designers have suggested that undesirable corner space be converted into a coat, storage, or broom closet, but this is seldom done. Shelves intersecting at outside corners do not pose nearly the same difficulty, and some revolving bookcases have exploited the geometry.

One especially awkward arrangement, which appears to be employed far more frequently than its usefulness would call for, is the corner shelf, a right-triangular form whose legs fit squarely against the two walls. To shelve books along the fore-edge of the shelf—the hypotenuse of the triangle—leaves them without solid support at the ends, unless appropriate bookends are used. This works, of course, but the wasted triangular spaces can drive some book owners to distraction. To shelve books along the legs of the triangle creates an incompatibility when the lines of volumes meet in the corner that is akin to the situation faced when bookcases intersect in a corner. In general, there is no good way to shelve books in a corner bookcase, yet they continue to be made, sold, bought, and installed. Some owners face the doubly vexing problem of how to shelve books in corner cases that have concave or worse profiles along their fore-edge, but the desire to use all available shelf space always wins out, and rectangular

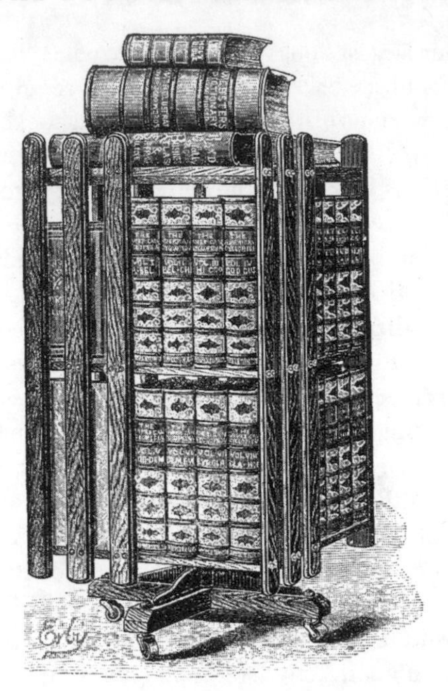

This Danner Revolving Book-Case was a late-Victorian device intended chiefly to hold reference books. In addition to rotating, it rolled on casters so that it could be moved about a library or study.

titles are fitted into triangular and worse holes. The improperly supported books are in the long run the losers. (The essayist Montaigne eliminated the problem of corners entirely by keeping his thousand-book library in a circular tower.)

Most books are kept on commonplace rectangular bookshelves, however, and while there are exceptions, those who do give a great deal of thought to their shelves also tend to care about how the books themselves are treated. Bookmarks are an especially touchy subject with some bibliophiles. The same depictions of medieval and Renaissance scholars that portray how they shelved books also show how they sometimes inserted slips of paper into them to mark a place or two. Dürer's *St. Jerome in His Cell* displays a book in the foreground closed and clasped around what appears to be a bookmark, but it is curiously inserted near the fore-edge rather than near the spine, which is where most modern readers would save their place. That not all Renaissance scholars followed Jerome's practice can be seen in Dürer's 1526 *Portrait of Erasmus,* in which volumes in the foreground contain bookmarks tucked in toward the spine.

Bookmarks could be placed as Dürer did Jerome's because when a book was closed tightly with its clasps there was little chance of the pressed bookmark becoming loose or falling between the pages. In modern times, it was the dislodging of loose slips of paper used as markers in his hymnal that inspired the engineer Art Fry to invent Post-it notes. These sticky but generally also removable tags have become indispensable to today's readers and scholars, who, like Fry, use them to mark passages in books. Unfortunately, the sticky stuff on the handy tags sometimes does not unstick easily from older books or magazines, thus tearing the page or lifting the type right off of it.

We did not have to wait for late-twentieth-century sticky notepaper to ruin books, however. Not surprisingly, Richard de Bury was sensitive to how some readers used stalks of straw and leaves and stems of grass to mark their places in books. Of the "headstrong youth lazily lounging over his studies," de Bury wrote,

> He distributes a multitude of straws, which he inserts to stick out in different places, so that the halm may remind him of what the memory cannot retain. These straws, because the book has no stomach to digest them, and no one takes them out, first distend the book from its wonted closing, and at length, being carelessly abandoned to oblivion go to decay. . . .
>
> Now the rain is over and gone, and the flowers have appeared in our land. Then the scholar we are speaking of, a neglecter rather than an inspector of books, will stuff his volume with violets, and primroses, with roses and quatrefoil. Then he will use his wet and perspiring hands to turn over the volumes; then he will thump the white vellum with gloves covered with all kinds of dust, and with his finger clad in long-used leather will hunt line by line through the page; then at the sting of the biting flea the sacred book is flung aside, and is hardly shut for another month, until it is so full of the dust that has found its way within, that it resists the effort to close it.

But such concerns for the welfare of books were not even new in the fourteenth century, as Vitruvius's worry over a poorly located library demonstrates: "In libraries with southern exposures the books are ruined by worms and dampness, because damp winds come up, which breed and nourish the worms, and destroy the books with mould, by spreading their damp breath over them." Yet as Adso wondered in *The Name of the Rose,* "What should be done? Stop reading, and only preserve?"

Preserving books is an admirable goal, and many libraries are engaged in important projects to arrest the deterioration of books printed on acid-based paper, which becomes brittle and disintegrates with age. But if only preservation were engaged in, libraries would be little more than museums. Just as a bookshelf does not make a library, so a line of books alone cannot make a bookshelf of any consequence. How many times have we seen cases in furniture stores filled with condensed books arranged like so much canned milk or window dressing or like throw pillows fluffed up on a sofa to soften up a customer for a sale? But while books alone do not make a bookshelf, they do make other books.

According to George Orwell, "People write the books they can't find on library shelves." But to write a book for the bookshelf is to take books down from the shelf and to use them in the way they were written to be used. They must sometimes necessarily be opened to the dirt and dust that blows in through the window and to the sweat and tears that come with toil in the fields of words. Sometimes grass and flowers and pens and pencils do have to be employed to hold more places than we have fingers. The books must be turned from page to page, backward and forward, sometimes frantically, in search of the way things have been, are, and will be. For all his prissiness, de Bury must have known, as I am so grateful that librarians today seem to know, that in the final analysis books are meant more than anything to be used. And I suspect that if push came to shove, de Bury would rather see the book soiled than never taken from the shelf, because he also wrote:

> For as in the writers of annals it is not difficult to see that the later writer always presupposes the earlier, without whom he could by no means relate the former times, so too we are to think of the authors of the sciences. For no man by himself has brought forth

> any science, since between the earliest students and those of the latter time we find intermediaries, ancient if they be compared with our own age, but modern if we think of the foundations of learning, and these men we consider the most learned.
>
> What would Vergil, the chief poet among the Latins, have achieved, if he had not despoiled Theocritus, Lucretius, and Homer, and had not ploughed with their heifer? What, unless again and again he had read somewhat of Parthenius and Pindar, whose eloquence he could by no means imitate? What would Sallust, Tully, Boethius, Macrobius, Lactantius, Martianus, and in short the whole troop of Latin writers, have done, if they had not seen the productions of Athens or the volumes of the Greeks?

Indeed, what would St. Jerome have done without the Greek and Hebrew scrolls that he threw about his bookshelves or the early Christian codices that he scattered at his feet, as if fodder for the lion and the lamb? As Boswell observed in his *Life of Dr. Johnson,* "A man will turn over half a library to make one book."

This is no less true today than it was at the time of St. Jerome or Dr. Johnson. When I was working on my first book, I used all sorts of other works to generate ideas, to find anecdotes, to verify facts, and generally to bolster my hypothesis that the concept of failure plays a central role in successful design. The cases in the study where I wrote that manuscript were already full of books, and so those I brought home from the library had to be placed somewhere else. At first they just accumulated in random piles on my desk, book upon book, in any but a vertical position, but soon these piles became dangerously high, were difficult to extract volumes from, and came to encroach on my working space. So I moved them from my desk to the floor, as institutional libraries are sometimes forced to do when the shelves simply cannot hold another book.

Instead of repiling the books one upon the other, however, I seem instinctively—at the time not yet having seen how libraries handle the situation—to have lined them up vertically before the bookcase as if the floor were a newfound shelf, but taking special precautions not to block the air vent that provided air conditioning to the room. The books stood more or less vertically on either side of the vent, the left side leaning left onto the wall's low baseboard, and the right side leaning right onto a cabinet fitted with toe molding. The books were in a convenient location for my work, and they were more or less arranged as they might be on a shelf.

But in the orientation that they were, I could not easily read their spines. Nevertheless, by the time the books got to the floor I had already become familiar not only with their insides but also with their outsides, and I could identify the book I needed by the color of its binding, the texture of the top edge of its paper, the thickness of its bulk, as if I were facing a medieval book press full of blank fore-edges.

I never did actually think of the floor before the bookcase as an additional shelf of any permanence, for when my writing project was done, I exposed the floor once more and left it bare. It remained the floor, which just happened to have once held a broken line of books set upon it. To this day I use the floor as a shelf only for library books, which I do not want to mix with my own, lest I forget to return something when it is due. Indeed, the floor is covered with books as I write this. But this time they are arranged spine up, so that I can read their titles more easily, though, again, because of the familiarity I have developed with them I seldom have to do so.

I have a new study now, and it has more real bookshelves, but they are pretty much full. It appears to be a law of nature that shelves, whether empty or full, attract books. These titles can be attracted from a considerable distance, like from the used bookstore in the next town or even from across the country or across the ocean. The attractive force between bookshelves and books is transmitted through the individual owned by the bookshelves or, for more formal collectors, through phone and fax to booksellers. When I travel, I find myself drawn into bookstores and to books I wonder if I will ever see again. Many of these volumes must be bought, of course, lest the opportunity to possess them be lost, and I have lugged inordinate numbers of titles through airports and squeezed overstuffed bags into undersized overhead bins (ill-formed bookshelves of a sort?).

Bookshelves have recently developed a new attractive force, one that operates through the Ethernet, the Internet, and the World Wide Web as Amazon.com and other virtual bookstores. So books can now be bought and transported via overnight delivery service without even being physically seen or touched by the bookshelf's custodian until the parcel containing them is opened. Home shelves are not virtual, however, and they fill up a lot faster than does a computer's hard disk. Bookshelves can also be emptied, of course, as they often are when those who love books get together and talk about them. Among the warmest memories I have of time spent with friends are those evenings when book after book was taken down from the shelf to find a favorite passage, to check an elusive fact, or to tease someone's memory.

Just as we clean up the glasses and dishes after our friends leave, however, so do we clean up the books, returning them to the shelves, there to rest in order to be handy for another party or another project. We need our books at the ready because, all too often, our memories are not as orderly as our bookshelves. As the Scotsman Andrew Lang wrote in his *Ballade of His Books*,

> *Here stand my books, line upon line*
> *They reach the roof, and row by row,*
> *They speak of faded tasks of mine,*
> *And things I did, but do not, know.*

When we work we often do so alone, of course, and the bookshelf and its books are the patient friends that surround us. We take them from the shelf to introduce them one to the other, comparing thoughts of sameness across generations and centuries, good-naturedly teasing one with the contradictory evidence of another. To write a book is to disrupt the bookshelf and threaten its serenity.

As I come to the end of this work, I see that the books are no longer line upon line, row by row on my shelves. I see a bit of the disorder of Jerome's study here, and I expect that for all his love of books, Richard de Bury's study was also in disarray as he came to the end of his *Philobiblon.* Although it is believed that entirely new texts began to appear with increasing frequency after the Middle Ages, in fact it is the unusual book even today that does not owe much to what is already on the bookshelf. When I am finished with this one, I will return all of my scattered volumes to a more or less orderly arrangement, allowing for the fact that there will in time be one more book to be accommodated. But where, and how, shall it be shelved?

Appendix

ORDER, ORDER

How can we arrange the books on our bookshelves? This question, like every question about order and design, is one to which there are many more answers than there are letters in the alphabet. Let us begin to count the ways—totally, half-so-, and not-so-seriously—in no particularly significant order, and without any claim of completeness or exhaustiveness:

1. BY AUTHOR'S LAST NAME. This alphabetical ordering is the plain-vanilla way of arranging books on bookshelves, but there are several problems one can encounter in doing so. For one, other than familiar standards like Bartlett's *Familiar Quotations* and Roget's *Thesaurus,* we tend not to remember the author of reference books, and might be hard pressed to locate our dictionary if it were not a Webster's or a Funk & Wagnalls.

Alphabetizing does have the advantage of grouping a particular author's works all in one place, but it has the obvious disadvantage of randomly distributing works on the same subject while juxtaposing those that make strange bedfellows. Susan Sontag is reported to have allowed that "it would set her teeth on edge to put Pynchon next to Plato." Thus the strict alphabetization by author of books, which at first would seem to be so logical and easy, soon reveals itself to be as fraught with frustrations and cautions as was the old library card catalog and is the modern online catalog.

In following a strict alphabetical order, one also has to make decisions as to whether to shelve O. Henry's books under O or H—or under P for the short-story writer's real name, William Sydney Porter—and then has to remember what was decided, and may even have to remember what O. Henry's real name was. One could, of course, incorporate directions into one's bookshelf, as an "O. Henry, *See* Porter, William Sydney" card used to do in many a card catalog. What form such a redirection should take on a bookshelf is itself problematic, for a card would likely be lost among the books. I have seen libraries put finding directions on book-sized blocks of wood that were shelved just like regular books, but too liberal a use of such a system would take up valuable space in a bookcase.

2. BY TITLE. We would not expect book titles to present the same problems to the alphabetizer as do the names of authors who use pen names, but difficulties can be encountered. When Donald Norman's *The Psychology of Everyday Things* was issued in paperback, its title was changed to *The Design of Everyday Things.* Thus, in a library that contained both titles, the same book would be shelved under both D and P. In a library that had only one, we might have to realize that the book existed under a different title than the one by which we know it.

Furthermore, many book titles are enigmatic (as Tracy Kidder's *The Soul of a New Machine,* about the creation of a computer) or allusive and thus elusive (as William H. Chafe's *Never Stop Running,* about the American liberal Allard Lowenstein), and we may or may not remember them when we are looking for the book among our titularly alphabetized volumes.

3. BY SUBJECT. This is an order that I tend to use, but subject can be a very elusive category. On one shelf I have a group of books dealing with design; on several other shelves I have books about bridges. Where do I put a book on the design of bridges? I have avoided having to make such Solomon-like decisions by choosing to have the bridge shelves directly beneath the design shelves. (Should I have located the bridges above?) The design of computers and computer software is another category of books that seems to be multiplying on my shelves. For some reason, I began to group these books together on the other end of the case, and so they are nowhere near the books on design generally, which they have much more to do with than the books about dams and dam disasters that they are now butted up against.

Booksellers like Barnes & Noble, whose names have been in alphabetical order since the business was founded in 1873, tend to use an order within an order. Books are grouped in broad subject categories and then generally alphabetized by author within that. For authors with multiple works within the same category, we can alphabetize those according to title, but then we will have really begun to complicate the ordering, for we will have alphabetized by title within an alphabetization by author within a subject category, which may or may not be in alphabetical order with respect to other subject categories in our collection. Though we may have become accustomed to such a pseudo-alphabetical arrangement, it is far from ideal and can be inconvenient when we are looking for new books by a favorite author who writes on several different subjects—or under different pseudonyms, though sometimes the genres are as different as the names and may as well be by entirely different authors.

4. BY SIZE. This is a popular way to arrange books, and I use it to a certain extent, but books about bridges, of which I have the several shelves mentioned above, come in a variety of sizes. There are the standard-sized books, generally referred to by librarians as octavos, such as David McCullough's *The Great Bridge*, his masterful story of the building of the Brooklyn Bridge. But like that structure's tall towers and long span, many books on bridges are either taller or wider than the standard size.

The larger formats of a lot of bridge books, loosely described as quartos, also fall into two categories, perhaps best described by the computer-printing terms, portrait mode and landscape mode. The terms are more or less self-explanatory, and the former allows for better showing the tall towers of bridges and the latter for showing their length. My main bridge bookshelf begins on the left end with a group of landscape-format books, which state departments of transportation seem to prefer for documenting their historic bridges. The smallest of these books is about 11 inches wide and so comes almost to the edge of my bookcase's 11⅛-inch-deep shelves (their irregular size is attributable to the fact that they have a piece of molding finishing their plywood core). Some other landscape bridge books are 12 inches wide and so overhang the bookshelf, which I find mildly annoying. The widest bridge book in my library, *Bridges* by Judith Dupré (which titularly joins *Bridges* by David J. Brown, *Bridges* by Fritz Leonhardt, *Bridges* by Steven A. Ostrow, and *Bridges* by Graeme and David Outerbridge, all classified as quartos in portrait mode), has a nonstandard format that measures 18 inches horizontally to its 7½ inches vertically, thus making it by far the widest (longest?) book I have in any category. If I were to shelve it in a conventional manner, this book would project almost 7 inches out from my bookshelf and would require a yellow flag to warn passersby. (Thus far I have kept the Dupré *Bridges* flat on a separate shelf, but I am not sure how long such shelfic extravagance can be maintained.)

After the landscape-type bridge books described above there begins a group of portrait ones, and these constitute the majority of titles on that shelf, for there are only a smattering of smaller octavos sandwiched in. Even though with tallness generally comes a slimming down in thickness, the mass of tall books makes the shelf look heavy. It does not seem to be sagging, however, because being only about 29 inches long, it is a relatively short span, especially by bridge-building standards. Nevertheless, there is a psychological heaviness to the shelf that is somewhat unsettling. I could, of course, relocate the larger books to one of the case's bottom shelves, where I do keep assorted tall volumes like dictionaries and coffee-table books. But that would make the bridge books, to which I refer frequently, less easily accessible.

Grouping books by size can thus present problems of appearance and accessibility. Indeed, the problem of what to do with large books generally has no easy solution. In the engineering library that I frequent, there is a group of large-format books, many having to do, not surprisingly, with bridges. Where these books are shelved vertically, they lean against each other in what seem to me to be precarious and straining ways, sure to lead to some form of biblioscoliosis. Where books have been removed, the gaps aggravate the situation, and many of the once-removed books are stacked horizontally atop the slanting pile, thus further aggravating the situation, apparently because it was not easy or possible for the patron or stack assistant to reinsert the book into the gap between the very heavy books. In the rare-book Dibner Library in the National Museum of American History at the Smithsonian Institution, the largest-format books are stored horizontally, as they were designed to be and as they are in the main stacks of the main library that I use at Duke. (Indeed, the engineering library has recently also installed wide horizontal shelves to take better care of its largest volumes.)

There are owners who wish to make a design or decorator statement with their books, by arranging them not with any intellectual order in mind—or even with any thoughts of ease of location down the line—but for maximum visual effect. A striking way of arranging one's books is to do so with a strict, no exceptions ordering by height, from the shortest book to the tallest, a scheme preserved to great effect in Samuel Pepys's library. This can produce an impressive effect in a wall of bookcases, and the effect can be even more impressive if the book order continues—in defiance of all good library practice—across all the top or bottom shelves, going from section to section of case, before continuing on to the next lower or higher shelves. Such an arrangement makes it imperative that all shelves be in alignment from section to section, of course, but that is considered normal good form anyway. Since the width of the books is totally ignored in this arrangement, there could be a ragged projection of volumes out from the bookcase, which it will be virtually impossible to prevent, especially if Dupré's *Bridges* is part of the collection. It will stick out from the wall the way Galileo's cantilever beam does from a Renaissance ruin in his *Dialogues Concerning Two New Sciences,* but being, at 7½ inches tall, among relatively short and generally small books, *Bridges* will look oddly unstable, like a loose piece of terra-cotta cornice on an old Chicago skyscraper, even if all the other books are aligned against the front edge of the shelf, as if to hide the board or the dust upon it as much as possible. The end of the line will also look odd if one owns Dupré's companion book, *Skyscrapers,* which is 18 inches tall but only 7½ inches wide. It will end up with the atlases and may look like mortar between bricks if the spines are not carefully aligned. Pepys solved the problem by shelving some of his books on their fore-edges. Doing so with the Dupré *Bridges* would juxtapose it with her *Skyscrapers.*

An alternative way to order books by size is to use width as a criterion for location. Viewed from straight on, the bookcase so arranged may look more or less randomly ordered. However, when seen from close up or from the side, there could be a distinct depth effect induced by the landscape books, likely projecting somewhat out from the facade like the odd stones in the Pont du Gard. For maximum effect when ordering books by width, they can be pushed all the way to the back wall of the bookcase, thus emphasizing the ordering principle and showing to greatest effect the fact that the books are indeed growing monotonically in width as one proceeds down the shelf. As in virtually all manners of ordering books, however, the Dupré books on skyscrapers and bridges will be as far removed from each other as two works by the same author are likely to be. Furthermore, the tall *Skyscrapers* will tower above its neighbors, and the wide *Bridges* will look like a holdout over air rights.

5. HORIZONTALLY. One of the most distressing situations to many a book lover and collector is to see books leaning against each other on shelves not filled in or on shelves from which a volume or two has been removed. The generally disheveled appearance of books leaning with their spines bent, their front and back covers skewed, their headbands aslant on their crowns, can be as unsettling as the sight of the Tower of Pisa. Substantial bookends can, of course, take care of the problem of a partially filled bookshelf, but how to keep books from leaning into a gap produced by extracting one of their neighbors is seldom addressed. One can, of course, fill the gap with another book of the approximate same size (preferably the exact size) as the volume removed, but if that title is taken from another section of the case the problem is merely relocated. One could keep a sacrificial shelf of books of graduated sizes at the ready to fill in where needed, but this would necessitate having a second-class shelf of books used for their brawn rather than their brains. If one has a case with a drawer or a cabinet at the bottom, one could keep in it blocks of wood, similar to the furniture of the old-time typesetter, from which could be selected the proper width to hold a book's place while it was

being used—a kind of quoin or bookmark for a bookcase. If these blocks were made of quality wood finished in appropriate stains, they could be very handsome indeed.

An alternative solution to the problem of books leaning, especially in bookcases with longish shelves, is to shelve one's books horizontally. Ashbel Green, my editor at Knopf, keeps the books in his office like this—matching the piles upon piles of manuscripts on his desk—and thus there are no misshapen volumes on the shelves. With the books arranged more or less alphabetically by the last names of Knopf authors, the spines of the books can be most easily read, with the exception of the odd book with a title of short words that has been set across rather than along the spine. Shelving books horizontally also addresses another frustration of book collectors, and that is the wasted space that invariably results when books of uneven height are shelved vertically. Horizontal shelving can indeed minimize the vertical voids in bookshelves, but some thought does have to be given to making sure the sizes of the books divide close to evenly into the length of the shelves or there will still be considerable wasted space.

6. BY COLOR. I once frequented the house of an engineering professor who had converted the family's dining room into an atelier of sorts. (When I joined the family for dinner one Thanksgiving, we ate off tables set up in the living room.) Homemade 7-foot-tall bookcases lined all the walls of the old dining room, thus wrapping around it in a striking way. Besides a few chairs scattered about the room, it held only one major piece of furniture, if it could be called that. In the far corner where two doorless walls conjoined, there was a bar-like structure built on what I recall to have been an elevated platform reminiscent of the study area in a Renaissance apartment. On the long broad surface that seemed intended to serve as a desk of sorts there were always scattered a great number of magazines, and these were an essential ingredient of the room, for they provided the raw materials of its unique decoration.

Every square inch of space above the bookcases—the walls and the ceiling—was covered with illustrations of all kinds cut from magazines long rendered unreadable. Anyone who visited the room was free to cut out a new picture and paste it on the walls or ceiling, but there was one rule that had to be followed: The picture had to be placed in an appropriate location according to its color, for the preexisting pictures were already so arranged, and the effect could not be lost. The room was in fact a spectrum of color, with the greatest area—on the ceiling and one wall—dominated by the flesh tones of bare-shouldered fashion models surrounding pin-up girls, who were most scantily clad. (This was in the

1960s, before nude photographs were commonly found in general-circulation magazines.) From this dominant color spread out in all directions the different colors of gowns worn by the fashion models. There was thus a blue section, a red section, a yellow section, and so forth, with the hues of one blending nicely into the tones of another. If my recollection serves me correctly, the scheme fell apart behind the counter-desk, as it had to somewhere in the room.

On occasion, usually when there were no newcomers to the distracting walls and ceiling, guests would gather in this room and discuss the books on the shelves. Since this was the house of an engineering professor, there were a good number of highly technical books, which, in color contrast to the paste-ups above the cases, were not brightened with dust jackets, and there were large runs of monotonically bound serials and journals. The order of the books had grown somewhat independently of the color motif, and it became a topic of conversation just how we might rearrange the books—no simple practical task with the number that there were—into a more logical or harmonious order. One way to move around the volumes became obvious to all who participated in this extra-credit homework exercise, and that was to carry the color scheme from the ceiling and walls down through the bookcases. There was, however, little hope of imposing such a design, because of the paucity of books with colorful bindings or dust jackets in the collection. Indeed, the presence of so many maroon-, gray-, and brown-bound volumes made the problem more than even a team of engineers could easily solve. To the best of my knowledge, the books never did get rearranged in that room.

But this is not to say that an impressive organization of books cannot be created by grouping them according to the dominant colors on their spines. Books with dust jackets usually work better than bare-spined cloth-bound titles, for all too many of those are black or dark-colored. Assuming one does have a colorful set of books, how to arrange them best might depend upon the general decor of the room in which the cases are located. If it does not have an evolutionary scheme as in the engineering professor's converted dining room, then it might make sense to pick up a wall color on the left and gradually evolve it into a different color to the right of the range of bookcases. Alternatively, one could go strictly by the spectrum and begin with red and end with violet. Leather-bound volumes make it possible to get a striking monochromatic effect, as has been done very effectively by some collectors.

7. BY HARDBACKS AND PAPERBACKS. Whether to intermix hard- and paperback books on bookshelves is a decision faced by everyone who wants to give order to a home library. If the types are intershelved, then

books are books within the decided-upon ordering scheme. The problem with juxtaposing paper- and hardbacks is that the square-backed former never look quite like the latter, the vast majority of which have the classic rounded spine of what most people think of as a real book. The effect is most striking when looking up at the higher shelves in a bookcase: the roundedness of the hardbacks will be clearly visible, and the rectangular profile of the paperbacks will be ever more obvious. In fact, if the paperbacks have not been handled with care during reading they are likely to have concave spines, contrasting them even more with the convexity of the hardcovers. And if the paperbacks have been read without any care to avoid cracking and folding their spines, then there can be little expectation that they will ever look good again, no matter how arranged on a bookshelf. Indeed, placing such books beside the handsomest leather-bound volumes can do nothing but diminish the appearance of the latter.

Segregating hard- and paperback books, on the other hand, creates a bookcase of singular dissonance. A shelf of paperbacks is likely to have a rigid look, with very little variation in height, especially if one tends to buy books of a particular kind. Mass-market paperbacks, for example, are generally of uniform size to fit readily into the racks beside supermarket and drugstore checkout lanes. Trade paperbacks also tend to have uniform sizes, though there is a bit more variation. In all cases with paperbacks, the principal visual distinction between the books on the shelf will be in the width of their spine, but this tends to be far less noticeable than the height differences or indifferences.

8. BY PUBLISHER. The Book Exchange in downtown Durham is famous for its large inventory of new and used books, including textbooks, and for the grouping of the bulk of its stock. Non-textbooks in the Book Ex are arranged according to publisher. The same scheme is followed for both hardbacks and paperbacks, but the two different kinds of binding are located in separate parts of the store. Within each publisher's section, books are further placed according to date of publication. For the hardbacks, this means that the most recently published books are at the end of the last shelf in a publisher's section; for paperbacks, the books are in numerical order according to the book's catalog number. Knowing the approximate date of publication of a book, and its publisher, of course, makes it relatively easy to find a title in the Book Ex, if one is familiar with where the various publishers are located, which is not always obvious or logical. New hardbacks are arranged in a way Melvil Dewey would have deplored, for a given publisher's line is literally that—a long horizontal band of shelves stretching around the room without regard for crossing vertical shelf supports. Not knowing the publisher

requires a trip to one of the many volumes of *Books in Print* that the Book Ex offers, but one must have a sense of the approximate publication date to know in which volumes to begin to look.

A private library could certainly be classified according to publisher, and this might provide an interesting study in different design philosophies. Many MIT Press books, for example, tend to be square-cased, which makes them flat-spined. Furthermore, the dust jackets of this publisher's books and the colors of its paperbacks tend to be of darkish hues. The masterful MIT colophon—comprising the minuscule sans-serif letters mitp executed with only seven equally spaced vertical lines, varying in height only to represent the ascender of the t and the descender of the p, and without the dot on the i or the cross on the t—would be another dominant design element that would tie the books together.

Another publisher's colophon, the borzoi of Alfred A. Knopf, would provide a striking curiosity in a bookcase arranged by publisher. Although the Russian wolfhound is always shown in a running position with its legs outstretched, there has been a great deal of subtle, and sometimes not so subtle, variation in the rendering of this mascot of sorts over the years. On my own books published by Knopf, the borzoi has been sketched in pencil, with all the strokes going in the same direction, as if made as a rubbing; the dog has been labeled as if it were a patent diagram for a useful thing; it has been reduced to curlicue lines, as if crafted out of wrought iron to be affixed to a nineteenth-century bridge; and it has been remade out of two triangles for the legs, an ellipse for the head, and what appears to be a segment of a circle for the tail to provide somewhat of an adventure in engineering. I will not be surprised to find on the dust jacket of this book a borzoi fashioned out of open books or running across a shelf. Collecting all of a library's Knopf books on the same bookshelf would provide a parade of borzois of striking individuality.

The foregoing ways to arrange books might be called public orderings. The principle behind a given grouping would be more or less evident to the casual bookshelf observer. There is another kind of scheme, which might be called a private ordering. Such an arrangement might be virtually impossible for anyone but the owner of the library to discern, though it might be puzzled out by the patient bookshelf sleuth.

9. BY READ/UNREAD BOOKS. Everyone has books that have not yet been read. What to do with them can be a problem. Some of us readers pile them up on tables or the floor beside a favorite reading chair, but this

can lead to an embarrassing reminder of how far behind we are in our reading. In extreme cases, the piles of accumulated unread books can present obstacles to getting to and from a chair and can hinder dusting and vacuuming. An alternative to segregating unread books is to integrate them fully into whatever ordering scheme we have adopted for our older, and presumably read, library. Thus we can shelve all books as they are acquired—through purchase, gifts, or whatever other means brings books to our floor and table—in the order that we have adopted for our bookcases, though this may require rearranging a lot of titles to make room for the newcomer. The intermixing of read and unread books, however, does present a problem when we are looking for a new book to read, for candidate books will tend to be spread randomly throughout our collection and so it would not be as convenient to consider the options as it is when we have a pile at our side. The stack also has the advantage of very likely being arranged in chronological order, so that if we wish to follow the "first acquired, first read" principle, we need only extract the bottom book from a pile—something that in itself may or may not be easy.

One way to maintain order among our unread books while at the same time keeping floor and table surfaces recognizable as such is to shelve unread books in a separate section of our case, say, at the beginning or the end of our read books. The end location is probably to be preferred, for most owners tend to acquire unread books at a faster rate than they read them, and so whatever shelf space might be allocated for new volumes at the beginning of a run of shelves, it is likely to be filled relatively quickly, thus necessitating the task owners seem to like least, reshelving an entire library to free up some space. (This is a potential problem with virtually any biblio-ordering, of course, but it can be expected to be especially acute when new titles are whisked to the front of the line.) Putting all our new books at the end of the shelving scheme would appear to solve the problem, but, as owners know, free shelving is always a temporary thing.

Books arranged in strict order of when they were read, with the unread grouped at the end in the order in which we plan to read them, can provide a privately revealing while publicly confusing order. We can see at a glance how our reading tastes evolved and our intellectual maturity developed from our school days to the present, and we can figure out where we expect to go in the future, or at least what books we plan to read on the journey. The ready grouping of books to be read certainly will make packing for a trip or vacation easy and mindless, though we might think twice if all the books heading the to-read group feature Dilbert and Garfield.

Another alternative way to arrange read and unread books is to shelve the former upside down within one's chosen ordering scheme. This would work fine for an all-American book collection, or at least one without any old British books, which characteristically had their titles imprinted on their spines or dust jackets in such a way that the title was read up from the right rather than the American way of down from the left. In a case with American and British books, one would have to know a volume's national origin to know whether it was read or unread.

An adventuresome way to shelve unread books among read ones is to hark back to the Middle Ages and shelve the unread spine inward, with the read spine out. This not only can add a bit of mystery to one's bookcase but also can make reading new books an adventure. Since it is unlikely that unread books will be recognized by their fore-edge, choosing a title to read will be a guessing game, unless its identity can be inferred by some overarching arrangement principle. Unless the books are ordered in some other way on the bookshelves—by category or by author, say—it is unlikely that readers will know exactly what unread book they are taking down from a case of spines and fore-edges. Choices might have to be made by thickness and texture, perhaps according to one's mood. However, what one might lose in choice one might gain in privacy, since the bookshelf snoop will certainly be frustrated. What one does with half-read books following this scheme is a problem, of course, but all ordering schemes ultimately lead to hard decisions. Perhaps the half-read could be shelved on their fore-edges.

10. BY STRICT ORDER OF ACQUISITION. Few if any of us seem to think about how to arrange our books until we have acquired a good number of them, too many in fact to remember the provenance or date of arrival of each of them reliably. Thus, arranging books in strict order of acquisition presents a guessing game with preexisting titles, but once started the scheme solves the problem entirely of what to do with new acquisitions. They are just added to the end of the last shelf. If we are of a mind to read the oldest unread book on our shelf, we need only begin at the beginning and pass our finger, as if a stick along a picket fence, along the spines until we come to the right title. If, on the other hand, we want to read the most recently acquired book, we need only take the last volume off the shelf.

Arranging one's books in a strict chronological order of when they were acquired can be very revealing of how one's taste in books has changed over the years. In a fairly typical library of a couple who met while they were graduate students, married, had children, saw the children through college, and were beginning to look ahead to retirement, the shelves might be heavy with poetry and philosophy in the beginning, but then might tend toward Dr. Spock (with or without being followed by a long run of juvenile titles, depending upon whether the children were still storing their childhood belongings at their parents' home), followed by a bunch of books on child psychology, adolescent psychology, and adult psychology, perhaps with some how-to and self-help guides intermixed; segueing into escapist literature, coffee-table books, and travel guides; with, if the children married late, bride's magazines and books on modern etiquette, investment and income-tax guides, and estate-planning manuals bringing up the rear. The shelf so arranged can be read like a book of life.

11. BY ORDER OF PUBLICATION. Another chronological ordering of books is by publication date, though this like all attempts to impose order on a set of artifacts can be fraught with decisions. What do we do with a second or later printing, or a reprinting by another publisher? Do we shelve the volume according to its own date of publication or according to when the first edition was published? Even a first edition can present some ambiguity of date, for the date on the title page, which is supposed to represent when the book was actually published, i.e., released to the world, might not agree with the date on the copyright page, which is the one usually used by bibliographers. A volume published in January 1998, for example, may carry that year on its title page but a copyright date of 1997. Subsequent reprintings of the book often carry the date of the reprinting on the title page. And then there are those annoying books that omit a date altogether, the kinds that appear in bibliographies with bracketed dates like [ca. 1968]. We can shelve such a book with the 1968s, of course, but the purist will always have the gnawing feeling that it is only approximately ordered, and it very well may be out of place by a year or two.

The difference between dating and arranging books by order of acquisition and order of publication can be very little for those who have generally bought nothing but new titles. However, for those who have, by want or necessity, developed a habit of purchasing used books, the difference can be astounding. When my wife and I were first married, I made a point of giving her a new first edition of poetry or fiction on our monthly anniversaries, and so the dates of publication and acquisition of these books would be virtually the same. However, when we decided that we wanted to fill out our library with the collected works of poets not represented, we began to look in used bookstores. Here the dates of publication and acquisition were years and decades apart, and so books that might be juxtaposed if ordered by publication date would be scattered here and there if shelved according to when we brought them home.

12. BY NUMBER OF PAGES. This ordering naturally tends to put thin books at the beginning and fat books at the end of a shelving arrangement. On first glance, the browser is likely to assume we have arranged our books according to thickness, but a closer examination would reveal that some thinner volumes are shelved after some fatter books because of the wide variation in the weights of paper used in manufacture.

A bookcase arranged according to number of pages does have the advantage of giving us a convenient way of choosing from among our shorter books when we are in a mood to do so, and from picking from among the longer works when we want to read without end.

13. ACCORDING TO THE DEWEY DECIMAL SYSTEM. Duke University and the University of Illinois at Urbana-Champaign are among the few large American research university libraries that still catalog and shelve books according to the Dewey decimal system. I was at the University of Texas at Austin when its library decided to convert from the Dewey to the Library of Congress system of classification, and it was an unsettling experience, for during the transition period a dual system of shelving was in effect, and one had to visit two places, with different local arrangements, to look for a book or to practice serendipity.

The Dewey decimal classification system was, of course, set down by the librarian Melvil Dewey, who as a student library assistant in 1873 applied a version of it to the Amherst College library. Dewey did not actually come up with the idea; his system was based upon W. T. Harris's scheme for arranging the books in the St. Louis Public Library. But it was Dewey who embraced the decimal, i.e., metric, system obsessively. Librarians, but few patrons, are aware of the fact that even the cards in card catalogs were decimal in that they were not the common 3- by 5-inch kind available in an ordinary stationery store but were metrically sized in centimeters (7.5 by 12.5), as were the heights of the books recorded on the cards. This meant, naturally, that the drawers in the card catalog furniture ideally were sized metrically also. Few card catalogs are maintained in the age of computers, of course, and the new San Francisco Library celebrated the demise of its main card catalog by decoupaging its cards, annotated by patrons poetic and prosaic alike, to the walls of the new building. The passing of the card catalog has been memorably noted by Nicholson Baker in a famous essay that appeared in *The New Yorker.*

One of the problems librarians of the late twentieth century have had with the Dewey decimal system is that the cataloging numbers—call numbers (named no doubt after the practice of calling for a book by the number designating its location in a book press)—can become interminably long. This is most easily seen at the beginning of the classification system, where the 001 designation was assigned by Dewey to books on information science. His assignment was made before the computer revolution, of course, and now books on computer science, artificial intelligence, and the like are all crowded into that one designation. In order to subdivide this category, books are assigned numbers like 001.53909, with further designators keyed to the first letter of the author's last name and the book's title, with the date of publication added to newer books. Thus, a full Dewey designation for Pamela McCorduck's *Machines Who Think: A Personal Inquiry into the History and Prospects of Artificial Intelligence* is: 001.53909 M131 M149 1979.

Since the majority of modern librarians seem not to be enamored of the nineteenth-century metric classification scheme, they are not likely to adopt it for their home libraries or recommend it to anyone else, especially since private libraries are more likely than their college or university counterparts to have their books grouped into a limited number of areas, such as bridges and design. Using the Dewey system to arrange such a library would mean that virtually all books would be classified under the same few whole and decimal numbers.

In Urbana, Illinois, the site of a distinguished library school, my wife and I once visited the home of a librarian and her husband and happened to be seated beside the bookcase in their living room. It was a modest-size case, and while the host and hostess were in the kitchen getting wine and cheese I naturally began to look at titles, but could discern no order that looked familiar. When the librarian returned, she saw my head cocked reading the spines of the books and informed me that the bookcase held but a small selection from their more extensive collection, which was distributed throughout the house. To give me a better sense of what books they owned, she brought me a box of index cards that constituted the catalog for the domestic collection, arranged not by the Dewey system but by one of her own devising—a truly private arrangement of books.

14. ACCORDING TO THE LIBRARY OF CONGRESS SYSTEM. The *Encyclopaedia Britannica* calls the LC system, as the Library of Congress system seems universally to be known, "an arbitrary rather than a logical or philosophical system of library organization developed during the reorganization of the U.S. national library; it consists of separate, mutually exclusive, special classifications, often having no connection save the accidental one of alphabetical notation." The encyclopedia goes on to point out that the LC system was based on an actual library of a million or so books rather than on the theory-based classification scheme of Dewey. The LC system does have the advantage of shorter classification designators, albeit in a completely alphanumeric form. Thus a book that

might have the designation 001.53909 M131 M149 1979 in the Dewey system would have the LC call number Q335 M23.

The big advantage of the LC system for public and private libraries alike is that every book now published in the U.S. carries its LC designation on the copyright page. Thus, one need not be a cataloger to assign a place on the shelf to any new American book. Because the LC system is "arbitrary rather than logical," the home library arranged according to it is unlikely to be recognized by the casual observer as being ordered in that way, unless the call numbers have been placed on the spines of the books.

15. BY ISBN. The International Standard Book Number (ISBN) is the number of ten or so digits that appears on the copyright page and often near the bars of the Universal Product Code (UPC) symbol on the back of virtually every book sold these days. The UPC system dates from the early 1970s—the grocery industry adopted the computer-based, IBM-developed way of expediting checkout and inventory tasks in 1973, and in 1974 a package of Wrigley's Spearmint gum was the first item sold via the attendant scanner technology—but books did not carry the rectilinear zebra markings until a decade or so later. Books purchased in the early-to-mid-1980s are likely to have their backs imprinted with a unique ISBN in squarish-looking numerals, and by the 1990s all books that publishers hoped to sell through retail markets also had the UPC bar code. This presented challenges to book-jacket designers, of course, and some have cleverly incorporated the stripes into their art work. Mostly, however, although the UPC-ISBN now defaces the backs of books, readers have come to ignore it or do not even consciously see it anymore.

That is not to say that the ISBN could not be used to order newer books at least, though it would be effectively like scheme No. 8 above. The ISBN for a book, e.g., 0-375-40041-9, always begins with a 0, so that is effectively ignored in arranging books. There follows a number representing the publisher. In this example, 375 identifies the publisher as Alfred A. Knopf. After the publisher's number is the individual book's number assigned by the publisher, usually in order of publication. Thus 40041 represents my 1997 book, *Remaking the World*. The last digit, here a 9, is known as a check digit. It is compared by the computer with a number that results from a formula applied to the previous digits. If the check digit does not agree with the calculation, the computer knows that there has been an error in reading the code, and it must be done again.

The corresponding UPC number for this book is 9-78035-400414. The UPC number is not exactly the same as the ISBN, but they clearly share a number of the same digits. For a book, the UPC always begins with a 9, thus identifying the category of the item. Bookshelves could, of course, be arranged by UPC number, but the order would not differ significantly from that arrived at according to ISBN.

16. BY PRICE. Although books arranged by price in constant dollars might be an order particularly difficult for the casual bookcase observer to decipher, it is usually more fun to arrange books by price in dollars current when the book was purchased. Beginning a run of books with trade paperbacks that once sold for 50 cents and decades-old hardback novels that sold for $1.95 when new can be an eye-opener. The bookcase arranged by purchase price can also be a source of interest for its general lack of correlation between size, quality, and price.

Of course, current value could also be used as an ordering index, and the owner who has both new and used books may want to make some judgments in shelving according to a price scale. Does a first edition of Robert Frost's *In the Clearing* purchased for $4.00 when it came out in 1962 (whole-dollar book pricing is not new!) get shelved with the four-dollar books or at the end of the line according to its current price in a rare-book catalog? And where does one put a 1979 first edition of William H. Gass's *The First Winter of My Married Life* that was inscribed and signed by the author on June 18, 1982, on the occasion of his and his family's being house guests and his presenting the book as a gift to his hosts? Although a presentation copy of the first edition that was limited to 275 numbered and 26 lettered copies signed by the author, the inscription certainly distinguishes the book and makes its current book market value an indeterminate multiple of its cost, which is technically zero. Does such a book belong at the beginning of the uppermost-left bookshelf, or should its location be changed with the changing market prices for Gass books?

17. ACCORDING TO NEW AND USED. Separating books into new and used purchases can also be employed in arranging a library. This is likely to result in a visually evident two-tiered system, however, for seldom do used books look as fresh and clean as new ones. This is also an ordering principle that is not likely to go unnoticed, unless perhaps one keeps new and used books in separate rooms.

18. BY ENJOYMENT. None of us enjoys every book equally. Some are pure pleasure to read, but others are drudgery. Our bookcases could reflect the range of emotion we feel by holding the most-enjoyed books in a privileged place and the least-enjoyed in an inferior place. A case so arranged might be in a constant state of flux, for books we remember so enjoying when first read do not always live up to our recollections when reread. It can also happen that many a book we disliked so much that we

put it away—in some dark and undesirable corner—unfinished becomes one of our favorites upon being given a second chance. However, what might prompt us to pick up such a book again may be as mysterious as what caused us to dislike it in the first place. The arrangement of books by likes and dislikes might encourage or discourage us to stoop to pick up a book we once consigned to the lowest shelf.

19. BY SENTIMENTAL VALUE. One of the most difficult decisions faced by owners is which volumes to discard when all the bookcases are overflowing and there is simply no more room in the studio apartment. Ordering books by sentimental value can prove to be a boon in such situations. If, for example, the books have been shelved with the least important ones on the bottom right of the last case in line, all one need do when it is spring book-cleaning time is to discard the contents of that shelf and open up all the shelves with the more important volumes, inserting in their proper place those newly sentimental titles that were awaiting a bookshelf berth. At the same time, of course, one might wish to rearrange one's bookshelves according to one's changed perceptions of sentimentality. The book given by a long-forgotten old beau or better-forgotten ex-spouse might be executed on the spot rather than moved to death row.

20. BY PROVENANCE. Although my own shelves are generally grouped by subject, I have several that appear to visitors to defy classification. These contain books grouped by where they came from. One shelf contains works signed by the author, another houses books sent by publishers seeking reviews. These latter titles have been sent to me largely because their authors, generally unknown to me, or editors equally unknown, think the books might have some interest for me, judging by what I have written. Though the range of subjects is broader than I can imagine ever making sense to the bookcase browser looking for an ordering principle, they all do have a connection to something I have written about, and so I am reluctant to discard them without having read them. In fact, those I have eventually gotten around to reading have proved to be rather interesting, suggesting to me that authors and editors do know what's good for me after all.

21. BY STILL MORE ESOTERIC ARRANGEMENTS. If one's object is truly to obfuscate the order of one's bookcase, there are countless other ways of doing so. Arrangements can be made by alphabetizing books according to subtitle, by author's first name, by opening sentence, by closing sentence, by third sentence, by antepenultimate sentence, accord-

ing to reverse-order spelling of last word in index, etc. Alternately, if one prefers a numerical ordering, one could array books by their number of words, or by the number of entries in the index. Clearly, each of these groupings has its disadvantages, not least of which is that one is not likely to want to count or alphabetize so tediously, and one is very unlikely to remember or know readily the necessary information to position or find a book in one's own library.

22. A NOTE ON CLOSET LIBRARIES. My wife and I once visited John Frederick Nims's apartment on Chicago's Lake Shore Drive with a commanding view of Lake Michigan stretching out like a calm ocean beyond. After some small talk, the poet offered us a drink and walked over to the bookcases in his living room, which were screened with a tight mesh that was very difficult to see behind. He opened the doors to the leftmost shelves, revealing not books but booze. There were books on some of the other shelves, but there was nowhere to be seen the working poetry library we expected, indeed knew him to have. We had been shown his study overlooking the lake, but we had not seen the library there either.

Perhaps sensing our confusion, John led us across the hall to a door, which he opened to reveal a large closet lined not with cedar but with narrow pine shelves jam-packed with the thin but scrappy volumes we had been looking for. As if a pantry for books, the closet held enough provisions for the mind, heart, and soul to keep the most voracious reader of poetry satisfied for an indefinite period of time. Had Chicago ever come under siege, this poet would not have wanted for reading material. (Although he must have at least dipped into most if not all of them, the volumes did not look the verse for wear.)

I believe the books were ordered alphabetically by author, which would be a natural arrangement. John pulled a few slim volumes off the shelf, as if to show us how easily he could put his hands on the work of any poet we could name. It was a striking display of an orderly arrangement of hundreds, if not thousands, of books in a single subject in a single closet in a single apartment on Lake Shore Drive, in the singular city of Chicago. How John could have squeezed in any books he acquired after our visit still remains a mystery to me. Perhaps book closets are like clothes closets, however, and no matter how full they appear, it is always possible to cram in one more jacket, even with its book in it. In fact, that seems to be the way it is with bookcases generally.

23. ON DUST JACKETS. Dust jackets can present an especially acute dilemma. Does one shelve books with the jackets on or off? To keep the jacket on conceals the book's binding, which has a color and character of

its own, but at the same time protects the binding from sun and soil. To remove the jacket separates interesting information, such as descriptive copy and photographs of authors, from the books. Some shelvers cut the jacket up and paste, tape, or just insert into the book what information they wish to keep, but I have never seen this done in a really neat way. Dust jackets can also be used to separate read from unread books, with the latter being shelved with their jackets on.

In any event, dust jackets do take up valuable space on the bookshelf, once calculated to be 2.5 percent of the total, which is space enough to accommodate a small public library for every million books. It is no wonder that crowded research libraries tend to discard jackets as just too extravagant a use of shelf space.

24. ON THE MARRIAGE OF TWO MINDS. Anne Fadiman begins her book *Ex Libris* with a charming essay, "Marrying Libraries," on how she and her husband finally intermingled their books after five years of preserving their individual, and seemingly incompatible, bookcase styles on opposite ends of their loft:

> His books commingled democratically, united under the all-inclusive flag of Literature. Some were vertical, some horizontal, and some actually placed *behind* others. Mine were balkanized by nationality and subject matter. Like most people with a high tolerance for clutter, George maintains a basic trust in three-dimensional objects. If he wants something, he believes it will present itself, and therefore it usually does. I, on the other hand, believe that books, maps, scissors, and Scotch tape dispensers are all unreliable vagrants, likely to take off for parts unknown unless strictly confined to quarters. My books, therefore, have always been rigidly regimented.

Her "French-garden" scheme won out over his "English-garden" one "on the theory that he could find his books if they were arranged like mine but I could never find mine if they were arranged like his." After a week of rearranging, their books were in "impeccable order," but as time passed disorder began to appear as books taken down, by him, to be read found their way back to different shelves. How familiar it all seems.

25. AND OTHER PROBLEMS. Other decisions that face the home librarian involve the wide variety of annoyances and opportunities for creative shelving, such as what to do with pamphlets, traditionally defined by the Stamp Act of 1712 as printed on no more than two sheets of paper. If each of these sheets were then folded three times, to produce

eight panels—octavo size—on each side of each sheet, the resulting pamphlet would have thirty-two pages. A more recent definition limits a pamphlet to something of no more than forty-nine pages. In any case, a pamphlet is always unbound, though it may have its pages stapled to paper covers.

When intermixed with books, pamphlets tend to get lost. Especially in a snugly packed bookcase, pamphlets can easily be pushed back when a regular book is reshelved. In the worst case, a pamphlet can be jammed against the back of the bookcase and be left unnoticed for years, to be discovered forgotten and disfigured only when a major rearranging or moving of the entire bookcase is undertaken. Grouping skinny pamphlets separate from bully books is not easily done, however, for a shelf of pamphlets in a home library presents an unattractive appearance—a shelf that looks like it holds trade catalogs in a warehouse.

Regardless of how books are grouped, they do furnish a room and so their appearance on the shelves deserves as much thought as does the placement of tables and chairs on the carpet, or the selection of pictures for the walls. The bookshelf without an orderly geometry of books can be as unsightly as the chair left askew at the dining room table or the picture frame hanging like a flag. The modern book is meant to stand tall on the shelf and not lean with its neighbors in a tug-of-war across empty space. When books cannot support themselves in an upright fashion, it is time to bring in the bookend.

Notes

Reference notes are keyed to quotations and phrases on the text pages indicated. Where successive quotes and information have come from the same source, only the first or most prominent occurrence is generally referenced. Full bibliographical citations for books and articles identified here only by author, or where multiple works by an author are cited, by author and short title, are given in the bibliography following these notes.

CHAPTER ONE. BOOKS ON BOOKSHELVES

PAGE:

7 **"The dust and silence"**: see Bartlett, p. 492

9 ***"I have a bookcase, which is what"***: ibid., p. 761

10 **"Can you grasp the sides"**: Wagner, p. 65
"To retrieve a book": Rogan, p. 90
"Never pull a book": Power, p. 129
"book ejection apparatus": U.S. Patent No. 4,050,754

12 **Alfred Kazin:** *New York Times,* June 6, 1998, obituary page, photo

14 **"too slender to bear a title"**: Fadiman, p. 139
"My brother and I": ibid., p. 124

16 **"best block"**: Penn, p. 118
patented in the 1870s: see ibid.

19 **"has enormous impact resistance"**: Ellis et al., p. 137
Charles Goodyear: alluded to in *Inventive Genius,* pp. 19–20

20 **"respected public figures"**: Gladwell, p. 68
"where there are bookshelves": ibid.
"Whenever we moved to a new house": Fadiman, p. 125
Thomas Jefferson's books: Brooks, p. [32]

21 **"The walls and shelves":** Smiles, p. 45
Admiral William J. Crowe, Jr.: Drucker and Lerner, pp. 16–17
David Macaulay: Marriott.

CHAPTER TWO. FROM SCROLLS TO CODICES

24 **On average, a scroll:** Clark, *Care of Books,* p. 27, n. 2
filled about a dozen rolls: Irwin, *Origins,* p. 117
"nearly three hundred running feet": Shailor, p. 6
"It is extraordinary": Kenyon, p. 66
"with a little practice": ibid.
25 **collected behind the scroll:** Clark, *Care of Books,* p. 28
26 **Keeping scrolls in a box:** ibid., p. 30
pigeonholes: ibid., pp. 33–35
28 **"Your men have made my library gay":** ibid., p. 33
"How can you excuse the man": quoted in Irwin, *Origins,* p. 45
29 **"Their old homes were restored":** de Bury, *Philobiblon,* p. 79
The codex evolved: Shailor, p. 8
30 **"by the beginning of the fourth century":** ibid.
31 **"occurs commonly in Cicero":** Clark, *Care of Books,* p. 37
a Britishism: American Library Association, entry "range"
32 **"the Roman classics were arranged":** Irwin, *Origins,* pp. 63–64
"small stones; large stone faces": Reed, p. 1
"The Book": ibid., p. 3
They were split open: see Kenyon, pp. 122–24
"well into the tenth and eleventh centuries": Shailor, p. 8.
"Papyrus was cheap and abundant": Irwin, *Origins,* pp. 115–16
33 **According to Pliny's *Natural History:*** Bradley, pp. 8–9
"made from sheep, goat": Shailor, p. 8
Vellum: Bradley, p. 8
"the skins of almost all the well-known domestic animals": ibid., p. 6
"some of the finest and thinnest": ibid., pp. 6–7
"one sheep yields no more than a single sheet": Irwin, *Origins,* p. 119
34 **"stitched sheets folded in concertina fashion":** Reed, p. 5
"rectangular" shape of animals: Weinstein, p. 29
beveled, cambered, or chamfered: Shailor, p. 59
35 **Lucretius's opening words:** Irwin, *Origins,* pp. 32–33

Les Miracles de Nostre Dame: Clark, *Care of Books,* Fig. 149
36 **drawing of Jean Mielot:** Weinstein, pp. 32–33
37 **Master of the Rolls:** Reed, p. 6

CHAPTER THREE. CHESTS, CLOISTERS, AND CARRELS

41 **"On the Monday after":** quoted in Clark, *Libraries,* p. 35
"The Librarian": quoted in Clark, *Care of Books,* p. 71
42 **"books were rare, and so was honesty":** Streeter, p. 3
Chests of this kind: Streeter, pp. 4, 117
43 **Another medieval book chest:** ibid., pp. 118–19
Simon, the abbot of St. Albans: see Clark, *Care of Books,* pp. 292–293
45 **"be lined inside with wood":** ibid., p. 71
47 **dates from the year 1232:** Irwin, *Origins,* p. 96
"curious wooden contrivances": ibid., p. 96
"tiny studies": Streeter, p. 5
48 **"plenty of complaints from scribes":** ibid., p. 95
"It was in the well-lighted carrel": Streeter, p. 5
49 **"where the monks write and study":** quoted in Clark, *Libraries,* p. 29
"From the great cloister": quoted in ibid., p. 30
occupants kept books behind closed doors: Irwin, *Origins,* p. 96
"be allowed to glance at books": quoted in Clark, *Care of Books,* p. 93
"places where private": Irwin, *Origins,* p. 96
"the appropriation of carrels": ibid.
52 **"the first library":** Riding, p. A18
"through heavy doors": ibid.
"the row of carrels": Streeter, p. 6
53 **"that curious book":** Clark, *Libraries,* p. 21
"an account by an eye-witness": Streeter, p. 5
"In the north syde of the Cloister": quoted in Clark, *Libraries,* pp. 21–22

CHAPTER FOUR. CHAINED TO THE DESK

56 **part of a sacristy:** Clark, *Care of Books,* pp. 84–85
"It is part of the precentor's duty": quoted in ibid., p. 69
57 **oldest known illustration of a book armarium:** ibid., pp. 40–41

58 **total space requirements:** Today, five hundred books might be shelved in a bookcase having a footprint of no more than about 10 square feet.

59 **"Books with clasps":** Power, p. 129

"books with carved bindings": ibid., p. 130

60 **"A chained book cannot be read":** Streeter, pp. xiii–xiv

62 **"Is it a desecration to assail":** ibid., p. 341

"Books of controversy": quoted in ibid.

63 **The process of unsecuring the chain rods:** see, e.g., Streeter, pp. 38–39

"there is no evidence": Streeter, p. 14

"assigned for division among the Fellows": ibid., p. 8

"those of which some are intended": quoted in ibid.

65 **pews at Hereford Cathedral:** ibid., pp. 104–108

"to convert the materials of the Bookcases": quoted in ibid., p. 280

66 **chained until the late eighteenth century:** ibid., p. 279

A classic example of a Gothic library: Clark, *Care of Books,* p. 153; see also Streeter, pp. 9–12

a latter stage in their evolutionary development: Streeter, p. 9

67 **The library at Zutphen is irregularly shaped:** Clark, *Care of Books,* p. 154

68 **The location of a library on an upper story:** see ibid., ch. III

a photograph of it survives: ibid., pp. 165–66 and fig. 64

Merton College Library: ibid., p. 179 and fig. 80

69 **Other English examples:** see ibid., chapter III

"that it might be the more safe": quoted in ibid., pp. 39–40

CHAPTER FIVE. THE PRESS OF BOOKS

74 **When Humphrey, Duke of Gloucester:** Clark, *Care of Books,* p. 171

75 **"should any student be poring over":** quoted in ibid., pp. 171–72

76 **termed the "stall system":** ibid., p. 172

as long as books were chained: ibid., pp. 264–65

"if the two halves of the desk": ibid., p. 172

combination of the lectern with the armarium: Streeter, pp. 44ff.

"they be packed so close": quoted in ibid., p. 45

Streeter dates the innovation: ibid., pp. 46–50

82 **"Probably the most often seen mistake"**: Ellis et al., p. 137
"the experience that shelves over 100 cm long": Dewey, p. 102
83 **A modern 36-inch bookshelf**: cf. Vogel, p. 60, but note that in his calculation Vogel assumes that the same total weight of books is placed on each shelf
86 **Cesna**: Clark, *Care of Books*, pp. 199–203
Trinity Hall: ibid., pp. 168–69
"the fairest of that University": quoted in Streeter, p. 69
"each side wall pierced": Clark, *Care of Books*, p. 249
88 **"originally 5 feet 6 inches high"**: ibid., pp. 250–251
standing lecterns that were a Cambridge tradition: Streeter, p. 69
89 **"stools were also provided"**: Clark, *Care of Books*, p. 251
91 **seat surface also served as a place on which to stand**: ibid.
92 **"the greater seats"**: ibid.
top shelf has no such verticals: see ibid., p. 250, fig. 110
"What was added to the lectern": Streeter, p. 46
93 **"the finest remaining British chained library"**: Hereford Cathedral brochure, "Mappa Mundi and Chained Library," ca. 1998
"Each book": Williams, p. 19
now known by American librarians as a "section": for bookstack nomenclature, see Henderson, "Bookstack Planning," p. 53; see also American Library Association
94 **"The shelf corresponds to the line"**: Dewey, p. 101
95 **"the notablest Library of Books"**: quoted in Irwin, *Origins*, p. 131
97 **"the public libraries of the Middle Ages"**: Clark, *Care of Books*, p. 245
"the whole system was swept away": ibid., p. 246
"exhibited itself in a very general destruction": ibid., p. 245
"upwards of eight hundred monasteries": ibid., p. 246
"to save any of the books": ibid.
"the buildings were pulled down": ibid.
Some books were sent by the shipload: ibid., p. 247
Manuscript pages were employed as endpapers: ibid.; see also Glaze
"to give us an imperfect notion": ibid., p. 246
98 **only three manuscripts were permitted to survive**: ibid., p. 248

98 **"had nothing superstitious about them"**: ibid., p. 247
"to sell, in the name of the University": quoted in ibid., p. 248
"nearly a century passed away": ibid.

CHAPTER SIX. STUDYING STUDIES

100 ***The Modern System of Naval Architecture:*** see Petroski, *Remaking the World,* pp. 139–143
102 **By around the year 1500**: Thornton, p. 62
A dialogue for Tudor schoolboys: quoted in ibid.
103 **increasingly ingenious devices**: for examples, see Clark, *Care of Books,* ch. IX
New College: Streeter, p. 7
107 **St. Jerome in his study**: see, e.g., Clark, *Care of Books,* figs. 140, 149, 153
Albrecht Dürer: *Encyclopaedia Britannica,* 15th ed. (1976), vol. 5, pp. 1085–1088
108 **damaged papyrus scrolls were being replaced**: Kenyon, p. 114
Benedetto Bonfigli: see Clark, *Care of Books,* p. 304
111 **"somewhat ungainly and archaic"**: *Encyclopaedia Britannica,* p. 1085
114 **"compelled to climb over his books"**: Irwin, *Origins,* p. 183
"who gathered books much as a squirrel gathers nuts": ibid.
"bibliocast and shoemaker": ibid.
"theaters of machines": Ferguson, p. 115
115 **"perhaps a thousand years before"**: Joseph Needham, p. 554
"the fact that Ramelli's was a vertical type": ibid., p. 555
"perfectly the preference of Western engineers": ibid., p. 547
"probably from the beginning": ibid., p. 554
116 **"A kind of square revolving bookcase"**: Lang, p. 37
"This is a beautiful and ingenious machine": Ramelli, ch. 188, p. 508
117 **"The wheel is . . . constructed"**: ibid.
118 **fifteenth-century illuminated manuscripts**: see, e.g. Thornton, figures on pp. 56–57
"for he was in the habit of fastening his queue": Streeter, pp. 15–16
119 **"clung to ancient fashions"**: Clark, *Care of Books,* p. 317
"before the seventeenth century": Irwin, *Origins,* p. 130
120 **Odorico Pillone**: Graham Pollard, p. 73, n. 2; see also Hobson

not totally accustomed to painting books: Hobson, p. 137

121 **"the spoliation of the monastaries":** Prideaux, p. 2

Twelve Centuries of Bookbindings: Paul Needham

"There are books of which": see Bartlett, p. 576

123 **"brought into harmony with that of the sides":** Prideaux, p. 53

124 **"the earliest tooled in gilt on the back":** Graham Pollard, p. 83

"following the fashion of the chained library": ibid., p. 73

library of the Spanish Escorial: conversation with Ron Druett, July 14, 1998; see also Graham Pollard, p. 73, n. 1

de Thou arranged it: Irwin, *Origins,* p. 176

126 **"dos à dos" bindings:** Paul Needham, pp. 281–284

"his books were growing numerous": quoted in Irwin, *Origins,* p. 172

"The truth is": see Allen, p. 39

The first cases were made for Pepys: *The Pepys Library,* p. 4

127 **"the placing as to heighth":** quoted in ibid., p. 7

"I think it will be as noble a closet": quoted in Irwin, *Origins,* p. 173

128 **"earliest known oak pedestal writing desk":** ibid., p. 197

CHAPTER SEVEN. UP AGAINST THE WALL

129 **This wall system:** see Clark, *Care of Books,* p. 267

130 **"The desks are 2 ft. 7 in. from the floor":** ibid., pp. 269–270

"the Escorial had a very definite effect": ibid., pp. 270–271

131 **"the room is not blocked with desks":** ibid., p. 272

"protected by wire-work": ibid., p. 271

132 **"but the severest penalties":** ibid., p. 272

modeled after those in the Escorial: Clark, *Libraries,* pp. 51–52

It was agreed in 1739 to add a gallery: Clark, *Care of Books,* pp. 272–273

"the finest that could be bought": ibid., p. 248

133 **the bookcases were sold in the 1550s:** ibid.

"the great east window": Craster, p. 6

These shadows are clearly present: see, e.g., Streeter, p. 73

134 **there are clear shadows cast:** see Craster, facing p. 6

required dress for any undergraduate: Craster, p. 3

asked to wear their academic dress: see <http://www.bodley.ox.ac.uk/guides/admisfrm.htm>

"in summertime, when honey-questing bees": Craster, p. 8

134 **painted on each ceiling panel:** ibid.
135 **bequeathed to the university by John Selden:** ibid.
136 **a series of shorter rods were used:** Streeter, p. 74
three or four thousand a year: Craster, p. 9
137 **"stowage for books":** ibid.
138 **"it ranks as the oldest public gallery in England":** ibid., p. 11
"all new octavos other than serials": ibid.
"blocking up its north windows": ibid.
"its north windows, like those of the north range": ibid.
"tall wire-fronted bookcases": ibid., p. 13
Francis Douce: ibid., pp. 15–16
139 **thus creating a cul-de-sac:** ibid., p. 16
"*à la moderne*": Clark, *Libraries,* p. 54
"rise high, and give place for the deskes": ibid., p. 52
"The disposition of the shelves": quoted in ibid., p. 53
142 **or countless other libraries:** see Snead & Company Iron Works, part 3, *passim*
reported having seen in England tall ladders: Dewey, p. 119
143 **"A common, light but strong ladder":** ibid., p. 120
144 **Michele Oka Doner:** Ellis et al., pp. 76–78
Congreave's Book Reacher: Dewey, pp. 120–21
"On a pole is a pair of metal jaws": ibid., p. 121
145 **"desiring to construct a door":** Jackson, p. 134

CHAPTER EIGHT. BOOKS AND BOOKSHOPS

146 ***incunabula:*** see, e.g., Carter
"fifteeners": Carter, p. 116
several hundred copies often constituted an edition: Power, p. 158
"a great book might be available": Weinstein, p. 5
147 **"could be available in thousands of copies":** Johnson, p. 163
more than one hundred thousand different books printed: Power, p. 158
Danse Macabre: Alfred Pollard, pp. 163–64
"the first picture-book ever made for children": Comenius, p. iii
Jan Amos Komensky: Graham Pollard, p. 92
"He keeps his stock in excellent order": quoted in Rivington, p. 4
"filing cabinets or bins": Graham Pollard, p. 92
150 **The labels on the bins:** ibid., pp. 92–93

"where the printer has printed the title": ibid., p. 91
"inside the cover to form a flap": ibid., pp. 91–92; for an illustration see plate I, facing p. 86
Printing on this leaf: Carter, pp. 108–109
"Never allow the binder": Power, p. 128
151 **"a careful study"**: Henderson, "Tiers," p. 382
"to fit snugly": ibid.
152 **"the common crror is waste of space"**: Dewey, p. 104
"The common O is only 15 cm": ibid.
"a quarto with a duodecimo": Power, p. 128
"backs were so narrow": Graham Pollard, p. 76
bind and sell their own books: Prideaux, p. 8
153 **done by master binders**: ibid., pp. 42–43
"rich, private collectors": ibid., p. 8
"Up, and by and by to my bookseller's": Pepys diary, January 18, 1664–1665, as quoted in Allen, p. 37
"Down to my chamber": ibid., February 5, 1665, as quoted in Graham Pollard, p. 74
"to Paul's churchyard": ibid., August 13, 1666
154 **"All the morning setting my books"**: ibid., February 2, 1667–1668
"so well ordered that his Footman": quoted in Douglas, p. 269 note.
"To the Strand": Pepys diary, February 8, 1667–1668, as quoted in Allen, p. 39
William Dugdale: see Douglas, ch. II, "The Grand Plagiary"
155 **"he had a special skill"**: ibid., p. 49
portrait engraved by Wenzel Hollar: see Douglas, frontispiece
157 **"in order to reduce the weight"**: Fadiman, p. 38
Napoleon Bonaparte: Brooks, p. [34]
Joshua Reynolds's famous 1775 portrait: see, e.g., Ellis et al., p. 40
Oxford University Press: Graham Pollard, p. 94, n. 1
158 **pasteboards**: Shailor, p. 59
rare-book stacks of one university library: University of Iowa
"As to the binding of books": quoted in Graham Pollard, p. 74
"the three passions": ibid.
159 **the customary manner of issuing an author's works**: ibid., p. 77
with agreed-upon prices published each year: ibid., p. 76

159 **priced to include a standard binding:** ibid., p. 74
imprint volume numbers: ibid., p. 77
160 **"a style of binding uniform for all copies":** ibid., p. 94
publisher itself began to bind: ibid., p. 76
161 **"seats were provided":** Rivington, p. 4
"Pepys found these shops useful": ibid.
162 **"Due to the reluctance of many readers":** Wheeler and Githens, p. 436
165 **"the last book":** see Lehmann-Haupt; see also Jacobson et al.
166 **promised for the 1998 Christmas season:** Lewis
Millennium Reader: see *New York Times*, October 8, 1998, national edition, p. D3

CHAPTER NINE. BOOKSTACK ENGINEERING

168 **"Since natural light":** Lydenberg, et al., p. 243
"reflect considerable light": Walter, p. 298
169 **"durability and economy in construction":** ibid.
Bibliothèque Sainte-Geneviève: Thompson, p. 71
"perhaps the last great flourish": *Encyclopaedia Britannica*, 15th ed. (1976), p. 864
"studious and curious persons": Wernick, p. 80
"narrow, dark, cold and damp": ibid.
170 **"engineering age":** *Encyclopaedia Britannica*, 15th ed. (1976), p. 864
"mouldings are particularly fine": Caygill, p. 36
"could not be used as a garden": ibid.
"to reduce the risk of fire spreading": Harris, p. 13
multilevel bookstacks: ibid., p. 14
171 **bookshelves in the Iron Library:** Thompson, p. 71
grate-like floors: see Snead & Company Iron Works, p. 67, fig. 105
172 **"turned the stack into a blast furnace":** Harris, p. 31
"a circular temple of marvelous dimensions": quoted in Caygill, p. 37
174 **The use of gas lighting was considered:** Harris, p. 23
Electric lighting was first attempted: ibid.
Further warmth was provided: Caygill, p. 36; see also Harris, p. 15
"a formal champagne breakfast": Harris, p. 17
175 **There was an attempt in 1920 to expand the capacity:** Harris, p. 28

inconvenient because entire bookcases had to be moved: Snead & Company Iron Works, p. 67, fig. 105
The hanging bookcases, moreover: Harris, p. 28
"Man builds no structure": see Bartlett, p. 713
"to celebrate the new millennium": see the flyer and brochure, *British Museum Great Court* and *Plans and Information, British Museum,* both available in October 1998 at the information desk in the British Museum
"longer than any other crane": ibid.

176 **"Into this were packed book ranges":** Snead & Company Iron Works, p. 11

177 **With the Copyright Act of 1870:** Thompson, p. 99
Bernard Richardson Green: Snead & Company Iron Works, frontispiece
The problem that faced: ibid., pp. 11–12

178 **"was so great an improvement":** Rider, *Compact Book Storage,* pp. 4–5
"library buildings, and their stacks": ibid., p. 5
"Green (Snead) standard": Lydenberg et al., p. 241

179 **the engineer might allow:** Snead & Company Iron Works, p. 83
250 pounds per square foot: see Boss, p. 397

180 **"great heat and more or less smoke":** Green, "Library Buildings," p. 55

181 **"Until very recently":** ibid., pp. 54–55; cf. Green in Snead & Company Iron Works, p. 118
"books, in fact": Green, in Snead & Company Iron Works, p. 118

183 **"not a building but a piece of furniture":** Green, "Library Buildings," p. 56
"open bar shelves": Snead & Company Iron Works, p. 30
"When shelves are to be *extra heavily* loaded": ibid., footnote to table

184 **"that scheme of shelving":** Green, "Library Buildings," p. 56
aisles took up 65 percent of the floor space: Thompson, p. 302
New York Public Library: Snead & Company Iron Works, p. 139
"simply drawn directly up through the floor": Green, "Planning and Construction," p. 681

185 **"a working laboratory for all kinds of people":** Macdonald, p. 1025

185 **"absolute control of air conditions"**: ibid., p. 972
"People's University": ibid., p. 1025
"We have windows, it is true": ibid., p. 973
186 **Bodleian Library constructed a new building**: Craster, p. 332
increasing calls to open bookstacks to patrons: ibid., p. 334
187 **"The near Stygian darkness"**: *Bindings* [newsletter of the Friends of the University of Iowa Libraries], Fall 1997, p. 8
"hand-lamps": Barwick, p. 147
188 **"sum people ar dizzy or unstedy on their feet"**: Dewey, p. 119
"a bracket step": ibid., p. 121
the unusually small man: see Petroski, *Beyond Engineering,* p. 21
190 **"Aisles lined with books"**: Dewey, p. 112
191 **it became the fashion in library architecture**: see, e.g., Macdonald, p. 973

CHAPTER TEN. SHELVES THAT MOVE

192 **"the astonishing growth"**: Rider, *The Scholar,* p. ix
193 **hard data in support of his observation**: ibid., pp. 3–8
Yale University Library: ibid., pp. 10–12
"why wood shelving is better than iron": Poole, p. 95
194 **"contrived to hold and hide indefinitely"**: Green, "Planning," p. 677
"It has been an axiom of library economy": Dewey, p. 109
"Rows upon rows of shelves": Henderson, "Tiers," p. 382
195 **"sum recommended that cloth or lether falls"**: Dewey, p. 114
Dewey also formulated a series of points: Dewey; see also Rider, *Compact Book Storage,* pp. 38–41
"architects design these miserable shelves": quoted in ibid., pp. 38–39
"Even in Wesleyan's relatively small library": ibid., p. 39, note
196 **A common way of providing for the adjustment**: Wheeler and Githens, p. 431
"ivory white, light greens and grays": Snead & Company, p. 37
197 **"industrial shelving, freestanding"**: Camp and Eckelman, pp. 775–777
"A domino effect": ibid., p. 777

"The card catalog toppled over": ibid., p. 777
"to head hight wherever we wish an aisle": Dewey, p. 100
198 **rolling or sliding bookcases**: Rider, *Compact Book Storage*, pp. 30–31; see also Garnett, "The Sliding-Press," p. 423
"the introduction of the principle": Garnett, "The Sliding-Press," p. 423
"because the British Museum aisles": Rider, *Compact Book Storage*, p. 31
subsequent stacks did not have: Snead & Company Iron Works, p. 67
"the process *could* be repeated": Dewey, p. 99
"although a person can pass": Snead & Company Iron Works, p. 67
199 **65 percent of the floor space**: Rider, *Compact Book Storage*, p. 8
install hinged bookcases: ibid., pp. 31–32
"Theorizers must not forget that this plan": Dewey, p. 100
200 **"may be lost for years"**: Rider, *Compact Book Storage*, p. 43
"The back of each shelf is fitted with a strip": Dewey, p. 104
201 **"It would seem to be at least possible"**: Rider, *Compact Book Storage*, p. 44
"slightly more easy to put it back": ibid.
202 **"Do not stand a book long on the fore-edge"**: Power, p. 129
203 **3⅓ sections of shelf space**: note that Rider, in *Compact Book Storage*, uses the term "tier" rather than "section"
Identifying books in the fore-edge down position: Rider, *Compact Book Storage*, pp. 49ff.
"this very proper query": ibid., p. 77
204 **University of Minnesota Library contract**: Walter, p. 297
"stuff into a single room": Fadiman, p. 142
"in a room well filled": Gladstone, p. 386
"the payment": ibid., p. 388
"with a passing anathema": ibid., p. 389
205 **"Now books want for"**: ibid., p. 391
"it contributes as a fastening": ibid., p. 392
"in illustration": ibid., p. 396, n. 12
"Let us suppose a room": ibid.
206 **"Why should not all the presses"**: Lymburn, p. 10
"versatile engineer": Locke, p. 554
"considerable freedom from dust": Lymburn, p. 10
208 **when the shelves are 84 percent filled**: Boss, pp. 396–97
209 **"a five-foot shelf"**: quoted in Brooks, p. [44]
"called attention to the urgent need": Elkins, p. 299

209 **"One who watches"**: quoted in Elkins, p. 300

210 **"depositories of dead books"**: Eliot, p. 55

"a cemetery of books": Stille, p. 43

"book-cemeteries": Gladstone, p. 396

"The student and the general reader": Eliot, p. 55

"sufficiently accessible if they could be delivered": Elkins, p. 305

opposed to classified storage: ibid., p. 306

211 **"It may be doubted"**: quoted in Metcalf, "The New England Deposit Library," p. 622

"enforced the point by writing the date": Elkins, p. 302

gift that enabled Gore Hall to be replaced: ibid., p. 311

New England Deposit Library: see Metcalf, "The New England Deposit Library," and Metcalf, "The New England Deposit Library after Thirteen Years"

212 **Midwest Storage Warehouse**: Rider, *Compact Book Storage*, pp. 15–16

automated storage and retrieval: Boss, pp. 398–99

as revolutionary as printing from moveable type: see McKee, p. 5

213 **reading material projected upon walls**: ibid.

"experience began soon to show": Thompson, p. 302

Nicholson Baker has chronicled: Baker

"new papyrus": quoted in McKee, p. 7

CD in allusion to BC and AD: ibid.

CHAPTER ELEVEN. THE CARE OF BOOKS

215 **"the delicate fragrant book-shelves"**: quoted in Lang, p. 4

"the teachers of Europe": ibid.

"There was once a bibliophile": ibid., pp. 35–36

216 **"readily into the pocket"**: ibid., p. 62

"of slack copyright": ibid., p. 109

"beautiful, but too small in type": ibid.

This skyline effect: see Petroski, "On Books, Bridges, and Durability"

Window shades: Ellis et al., p. 199

"won't let his wife raise the blinds": Fadiman, p. 43

217 **"buys at least two copies"**: ibid.

"castlelike folly": ibid., p. 37

"lined in billiard-table baize": ibid., p. 39

Bern Dibner: see Petroski, "From Connections to Collections," p. 417

218 **"really wanted to be a reader"**: Ellis et al., p. 173
"made room for books": ibid., p. 193
220 **"great injury is done so often"**: de Bury, *Philobiblon,* p. 155
"being shown by the horrified Superintendent": Harris, p. 23
breakfast served off the catalog desks: ibid., p. 17
"the race of scholars": de Bury, *Philobiblon,* p. 155
"You may happen to see": ibid., p. 157
221 **"a very handsome dress"**: Eco, p. 185
"Its pages crumble": ibid.
223 **"was so crowded with books"**: Gussow, p. B8
224 **One book accumulator's wife**: Ellis et al., p. 137
"What are you doing?": Gussow, p. B8
"mutilators of collections": quoted in Ellis et al., p. 194
"every hole in the shelf a crater": ibid.
226 **Montaigne**: see Brooks, p. [1]
227 **Post-it notes**: see, e.g., Petroski, *The Evolution of Useful Things,* pp. 84–86
"headstrong youth lazily lounging": de Bury, *Philobiblon,* p. 157
"He distributes a multitude of straws": ibid., pp. 157, 159
228 **"In libraries with southern exposures"**: Vitruvius, p. 181
"What should be done?": Eco, p. 185
George Orwell: quoted in Brooks, p. [1]
"For as in the writers of annals": de Bury, *Philobiblon,* p. 111
231 ***"Here stand my books":*** see Bartlett, p. 725
entirely new texts began to appear: Weinstein, p. 30

APPENDIX: ORDER, ORDER

233 **"it would set her teeth on edge"**: Fadiman, p. 6
245 **as a student library assistant**: Rider, *Melvil Dewey,* p. 30
famous essay: Baker
247 **package of Wrigley's Spearmint gum**: *Newsweek,* special issue, Winter 1997–98, p. 30; see also Leibowitz, p. 138
250 **Dust jackets**: see Petroski, *Beyond Engineering,* pp. 149–156; see also *New York Times Book Review,* May 18, 1986, p. 21
251 **once calculated**: see Petroski, *Beyond Engineering,* p. 151
"His books commingled democratically": Fadiman, pp. 4–5

Bibliography

This book began with a question: From where did the bookshelf and bookcase come? The question soon turned into a hypothesis: The bookcase evolved, as I believe all artifacts do, in response to real and perceived problems with existing technology. In the case of the bookshelf, this meant shortcomings in the way in which books were stored.

Testing such a hypothesis can begin, at least partially, in books, of course. Thus, my reading took the form of a search for confirmations or refutations of the idea. My interest in the evolution of the bookcase led me to look for scholarship, artifacts, and illustrations of all kinds to test the hypothesis, and I found it confirmed and anticipated in many of the items listed in this bibliography, the most helpful of which have been John Willis Clark's *The Care of Books* and Burnett Hillman Streeter's *The Chained Library.*

Though hypotheses about artifacts of technology, like all hypotheses, may be tested and verified, they can never be proved in any mathematical sense, of course. If there are additional sources that might have provided counterexamples, thus disproving what I have set out to demonstrate, I have not discovered them in my admittedly incomplete survey, recorded here, of the literature on books, libraries, and their furniture.

Allen, Edward Frank, ed. *Red-Letter Days of Samuel Pepys.* New York: Sturgis and Walton, 1910.

American Library Association. *The ALA Glossary of Library and Information Science.* Chicago: American Library Association, 1983.

Baker, Nicholson. "Discards," *The New Yorker,* April 4, 1994, pp. 64–86.

Barker, Nicolas. *Treasures from the Libraries of National Trust Country Houses.* New York: Royal Oak Foundation and Grolier Club, 1999.

Bartlett, John. *Familiar Quotations.* 13th ed. Boston: Little, Brown, 1955.

Barwick, G. F. *The Reading Room of the British Museum.* London: Ernest Benn, 1929.

Beinecke Library. *The Beinecke Rare Book & Manuscript Library: A Guide to the Collections.* New Haven: Yale University Press, 1994.

Birley, Robert. "The History of Eton College Library," *Library* 11 (December 1956): 231–261.

Boss, Richard W. "Space Conserving Technologies," *Library Technology Reports* 31 (July–August 1995): 389–483.

Bradley, John. *Illuminated Manuscripts.* London: Bracken Books, 1996.

Bright, Franklyn F. *Planning for a Movable Compact Shelving System.* Chicago: American Library Association, 1991.

Brooks, Marshall. *A Brief Illustrated History of the Bookshelf: With an Essay Which Pertains to the Subject.* Delhi, N.Y.: Birch Brook Press, 1998.

Bryant, Arthur. *Samuel Pepys: The Man in the Making.* New edition. London: Collins, 1947.

Bury, Richard de. *The Love of Books: The Philobiblon.* Translated by E. C. Thomas. New York: Barse & Hopkins, 1903.

———. *Philobiblon.* Translated by E. C. Thomas. Oxford: Basil Blackwell, 1960.

Calkins, Robert G. *Illuminated Books of the Middle Ages.* Ithaca, N.Y.: Cornell University Press, 1983.

Camp, John F., and Carl A. Eckelman. "Library Bookstacks: An Overview with Test Reports on Bracket Shelving," *Library Technical Reports* 26 (November–December 1990): 757–894.

Carter, John. *ABC for Book-Collectors.* Fourth edition. New York: Alfred A. Knopf, 1966.

Caygill, Marjorie. *The Story of the British Museum.* Second edition. London: British Museum Press, 1992.

[Clare College.] *Clare College, 1326–1926: University Hall, 1326–1346; Clare Hall, 1346–1856.* Volume II. Cambridge: Clare College, 1930.

Clark, J. W. *Libraries in the Medieval and Renaissance Periods.* Chicago: Argonaut, 1968. Reprint of 1894 edition.

Clark, John Willis. *The Care of Books: An Essay on the Development of Libraries and Their Fittings, from the Earliest Times to the End of the Eighteenth Century.* Cambridge: University Press, 1901.

Comenius, John Amos. *The Orbis Pictus.* Syracuse, N.Y.: C. W. Bardeen, 1887.

Communications of the ACM. Special issue on digital libraries. April 1998.

Condit, Carl W. *American Building Art: The Nineteenth Century.* New York: Oxford University Press, 1960.

Cooper, Gail. *Air-Conditioning America: Engineers and the Controlled Environment, 1900–1960.* Baltimore, Md.: Johns Hopkins University Press, 1998.

Craster, Sir Edmund. *History of the Bodleian Library, 1845–1945.* Oxford: Clarendon Press, 1952.

Daumas, Maurice, ed. *A History of Technology & Invention: Progress through the Ages. Volume I: The Origins of Technological Civilization.* Translated by Eileen B. Hennessy. New York: Crown Publishers, 1969.

Dawe, Grosvenor. *Melvil Dewey: Seer, Inspirer, Doer, 1851–1931.* Essex Co., N.Y.: Lake Placid Club, 1932.

[Dewey, Melvil. "Notes on Library Shelving."] *Library Notes* 2 (September 1887): 99–122.

Diringer, David. *The Book Before Printing: Ancient, Medieval and Oriental.* New York: Dover Publications, 1982.

Douglas, David C. *English Scholars, 1660–1730.* Second edition. London: Eyre & Spottiswoode, 1951.

Drucker, Hal, and Sid Lerner. *From the Desk of.* San Diego: Harcourt Brace Jovanovich, 1989.

Eco, Umberto. *The Name of the Rose.* Translated by William Weaver. San Diego: Harcourt Brace, 1984.

Eliot, Charles William. "The Division of a Library into Books in Use, and Books Not in Use, with Different Storage Methods for the Two Classes," *Library Journal* 27 (1902, Magnolia Conference Supplement): 51–56, 256–257.

Elkins, Kimball C. "President Eliot and the Storage of 'Dead' Books," *Harvard Library Bulletin* 8 (1954): 299–312.

Ellis, Estelle, Caroline Seebohm, and Christopher Simon Sykes. *At Home with Books: How Booklovers Live with and Care for Their Libraries.* New York: Carol Southern Books, 1995.

Engler, Nick. *Desks and Bookcases.* Emmaus, Pa.: Rodale Press, 1990.

Esdaile, Arundell. *The British Museum Library: A Short History and Survey.* London: Allen & Unwin, 1946.

Fadiman, Anne. *Ex Libris: Confessions of a Common Reader.* New York: Farrar, Straus and Giroux, 1998.

Ferguson, Eugene S. *Engineering and the Mind's Eye.* Cambridge, Mass.: MIT Press, 1992.

Garnett, Richard. "The Sliding-Press at the British Museum," *Library Journal* 17 (October 1892): 422–424.

———. "On the Provision of Additional Space in Libraries," *Library* 7 (1895): 11–17.

———. "New Book Press at the British Museum," *Library Notes* 2 (September 1897): 97–99.

Gawrecki, Drahoslav. *Compact Library Shelving.* Translated by Stanislav Rehak. Chicago: American Library Association, 1968.

Gayle, Addison, Jr. *Oak and Ivy: A Biography of Paul Laurence Dunbar.* Garden City, N.Y.: Doubleday, 1971.

Gladstone, W. E. "On Books and the Housing of Them," *Nineteenth Century* XXVII (1890): 384–396.

Gladwell, Malcolm. "The Spin Myth," *The New Yorker,* July 6, 1998: 66–73.

Glaze, Florence Eliza. "Hidden for All to See," *Duke University Libraries* 12 (Fall 1998): 2–7.

Green, Bernard R. "Planning and Construction of Library Buildings," *Library Journal* 25 (1900): 677–683.

———. "Library Buildings and Book Stacks," *Library Journal* 31 (1906): 52–56.

Griliches, Diane Asseo. *Library: The Drama Within.* Albuquerque: University of New Mexico Press, 1996.

Gussow, Mel. "$8 Million Literary Trove Given to Morgan Library," *New York Times,* February 23, 1998, national edition, pp. B1, B8.

Hall, Bert S. "A Revolving Bookcase by Agostino Ramelli," *Technology and Culture* 11 (1970): 389–400.

Harris, P. R. *The Reading Room.* London: The British Library, 1979.

Henderson, Robert W. "Tiers, Books and Stacks," *Library Journal* 59 (1934): 382–383.

———. "The Cubook: A Suggested Unit for Bookstack Measurement," *Library Journal* 59 (November 15, 1934): 865–868.

———. "Bookstack Planning with the Cubook," *Library Journal* 61 (January 15, 1936): 52–54.

Hobson, A. R. A. "The Pillone Library," *Book Collector* 7 (1958): 28–37.

International Correspondence Schools. *The Building Trades Pocketbook: A Handy Manual of Reference on Building Construction, Including Structural Design, Masonry, Bricklaying, Carpentry, Joinery, Roofing, Plastering, Painting, Plumbing, Lighting, Heating, and Ventilation.* Scranton, Pa.: The Colliery Engineer Co., 1899.

Inventive Genius. Library of Curious and Unusual Facts. Alexandria, Va.: Time-Life Books, 1991.

Irwin, Raymond. *The English Library: Sources and History.* London: George Allen & Unwin, 1966. Revised and enlarged edition of *The Origins of the English Library.*

———. *The Heritage of the English Library.* New York: Hafner, 1964.

———. *The Origins of the English Library.* London: George Allen & Unwin, 1958.

Jackson, Holbrook. *The Anatomy of Bibliomania.* New York: Farrar, Straus, 1950. Reissued as *The Book about Books: The Anatomy of Bibliomania.* New York: Avenel Books, 1981.

Jacobson, J., et al. "The Last Book," *IBM Systems Journal* 36 (1997): 457–463.

Jarrell, Randall. *Jerome: The Biography of a Poem.* New York: Grossman, 1971.

Jenner, Henry. "Moveable Presses in the British Museum," *Library Chronicle* 4 (1887): 88–90.

Johnson, Elmer D. *History of Libraries in the Western World.* Second edition. Metuchen, N.J.: Scarecrow Press, 1970.

Kaplan, Louis. "Shelving." In *The State of the Library Art.* Vol. 3, Part 2. Ralph R. Shaw, ed. New Brunswick, N.J.: Rutgers Graduate School of Library Service, 1960.

Kenyon, Frederic G. *Books and Readers in Ancient Greece and Rome.* Oxford: Clarendon Press, 1932.

Kimber, Richard T. *Automation in Libraries.* Oxford: Pergamon Press, 1968.

Lang, Andrew. *The Library.* London: Macmillan, 1881.

Latham, Robert, ed. *The Shorter Pepys.* Berkeley: University of California Press, 1985.

Lehmann-Haupt, Christopher. "Creating 'the Last Book' to Hold All the Others," *New York Times,* April 8, 1998, national edition, pp. B1–B2.

Leibowitz, Ed. "Bar Codes: Reading Between the Lines," *Smithsonian,* February 1999: pp. 130–132, 134–146.

Levarie, Norma. *The Art & History of Books.* New Castle, Del., and London: Oak Knoll Press and the British Library, 1995.

Lewis, Peter H. "Taking on New Forms, Electronic Books Turn a Page," *New York Times,* July 2, 1998, national edition, pp. G1–G7.

Library of Congress. *Report of the Librarian of Congress and Report of the Superintendent of the Library Building and Grounds.* Washington, D.C.: Government Printing Office, 1909.

Locke, George H. "Toronto Method of Book Storage," *Library Journal* 56 (June 15, 1931): 554.

Lydenberg, H. M., et al. "Bookstacks: The Librarians' Viewpoint," *Library Journal* 41 (1916): 238–244.

Lymburn, John. "Suspended Iron Presses for Book Accommodation in Large Libraries," *Library Journal* 18 (January 1893): 10.

Macdonald, Angus Snead. "A Library of the Future, Part I," *Library Journal* 58: 971–975.

———. "A Library of the Future, Part II," *Library Journal* 58: 1023–1025.

Manguel, Alberto. *A History of Reading.* New York: Penguin Books, 1997.

Marriott, Michel. "For Mundane and Exotic, a Guide," *New York Times,* March 12, 1998, national edition, p. D3.

McKee, Eugenia Vieth. *The Diffusion of CD-ROM as a Text Information Storage Technology for Libraries: A Comparative Study.* Ph.D. diss., Texas Woman's University, Denton, Texas, 1989.

McKerrow, Ronald B. *An Introduction to Bibliography for Literary Students.* Oxford: Clarendon Press, 1928.

Metcalf, Keyes D. "The New England Deposit Library," *Library Quarterly* 12 (1942): 622–628.

———. "The New England Deposit Library after Thirteen Years," *Harvard Library Bulletin* 8 (1954): 313–322.

———. *Planning Academic and Research Library Buildings.* New York: McGraw-Hill, 1965.

Morley, Christopher. *Parnassus on Wheels.* New York: Lippincott, 1917.

Naudeus, Gabriel. *Instructions Concerning Erecting of a Library: Presented to My Lord The President de Mesme.* Translated by Jo. Evelyn. Cambridge, Mass.: Houghton, Mifflin, 1903.

Needham, Joseph. *Science and Civilisation in China. Volume 4: Physics and Physical Technology. Part II: Mechanical Engineering.* Cambridge: Cambridge University Press, 1965.

Needham, Paul. *Twelve Centuries of Bookbindings, 400–1600.* New York and London: The Pierpont Morgan Library and Oxford University Press, 1979.

Nixon, Howard M., and William A. Jackson. "English Seventeenth-Century Travelling Libraries," *Transactions of the Cambridge Bibliographical Society* 7 (1977/1980): 294–322.

Ollé, James G. *Library History: An Examination Guidebook.* Second edition. London: Clive Bingley, 1971.

Orne, Jerrold. "Storage Warehouses." In *The State of the Library Art.* Vol. 3, Part 2. Ralph R. Shaw, ed. New Brunswick, N.J.: Rutgers Graduate School of Library Science, 1960.

"*Other* Side of the Counter, The." " 'A Plea for Liberty' to Readers to Help Themselves," *The Library,* first series, 4 (1892): 302–305.

Paintin, Elaine M. *The King's Library.* [London]: The British Library, n.d.

Penn, Arthur. *The Home Library.* New York: D. Appleton, 1883.

The Pepys Library. [Cambridge: Magdalene College.] n.d.

Petroski, Henry. *Beyond Engineering: Essays and Other Attempts to Figure Without Equations.* New York: St. Martin's Press, 1986.

———. *The Pencil: A History of Design and Circumstance.* New York: Alfred A. Knopf, 1990.

———. *The Evolution of Useful Things.* New York: Alfred A. Knopf, 1992.

———. "On Books, Bridges, and Durability," *Harvard Design Magazine,* Fall 1997: 19–21.

———. *Remaking the World: Adventures in Engineering.* New York: Alfred A. Knopf, 1997.

———. "From Connections to Collections," *American Scientist,* September–October 1998: 416–420.

Pollard, Alfred W. *Early Illustrated Books: A History of the Decoration and Illustration of Books in the 15th and 16th Centuries.* London: Kegan Paul, Trench, Trübner: 1893.

Pollard, Graham. "Changes in the Style of Bookbinding, 1550–1830," *Library,* 11 (1956): 71–94.

Poole, Wm. F. "Why Wood Shelving Is Better than Iron," *Library Notes* 2 (September 1887): 95–97.

Powell, Anthony. *Books Do Furnish a Room.* Boston: Little, Brown, 1971.

Power, John. *A Handy-book About Books, for Book-lovers, Book-buyers, and Book-sellers.* London: John Wilson, 1870.

"P.P.C.R." "The New Building at the British Museum," *Mechanics' Magazine, Museum Register, Journal, and Gazette,* March 1837: 454–460.

Prideaux, S. T. *An Historical Sketch of Bookbinding.* London: Lawrence & Bullen, 1893.

Puccio, Joseph. "Managing 10 Million Volumes and Counting: Collections Management Oversees 257 Miles of Shelves," *Library of Congress Information Bulletin,* May 4, 1992: 189–194.

Ramelli, Agostino. *The Various and Ingenious Machines of Agostino Ramelli (1588).* Translated and edited by Martha Teach Gnudi and Eugene S. Ferguson. Baltimore: Johns Hopkins University Press, 1976.

Ranz, Jim. *The Printed Book Catalogue in American Libraries: 1723–1900.* Chicago: American Library Association, 1964.

Reed, R. *Ancient Skins, Parchments and Leathers.* London: Seminar Press, 1972.

Rider, Fremont. *Melvil Dewey.* Chicago: American Library Association, 1944.

———. *The Scholar and the Future of the Research Library: A Problem and Its Solution.* New York: Hadham Press, 1944.

———. *Compact Book Storage: Some Suggestions Toward a New Methodology for the Shelving of Less Used Research Materials.* New York: Hadham Press, 1949.

Riding, Alan. "Mitterrand's Last Whim Struggles with Reality," *New York Times*, November 7, 1998, national edition, p. A18.

Rivington, Charles A. *Pepys and the Booksellers.* York: Sessions Book Trust, 1992.

Rogan, Helen. "Organizing," *Martha Stewart Living*, February 1999: 86, 88, 90.

Russell, John Scott. *The Modern System of Naval Architecture.* Three volumes. London: Day and Son, 1865.

Shailor, Barbara A. *The Medieval Book: Illustrated from the Beinecke Rare Book and Manuscript Library.* Toronto: University of Toronto Press, 1991.

Shepherd, Jane Bushnell. *Miss Antoinette Turner's Store: And Other Reminiscent Sketches.* New Haven: Tuttle, Morehouse & Taylor, 1929.

Smiles, Samuel, ed. *James Nasmyth, Engineer: An Autobiography.* London: John Murray, 1885.

Smith, Alexander. *Dreamthorp: A Book of Essays Written in the Country.* Garden City, N.Y.: Doubleday, Doran, 1934.

Snead & Company. *Snead Bookstacks.* New York: Snead, 1940.

Snead & Company Iron Works. *Library Planning, Bookstacks and Shelving.* Jersey City, N.J.: Snead, 1915.

Steinberg, S. H. *Five Hundred Years of Printing.* New York: Criterion Books, 1959.

Stevenson, Robert Louis. *Familiar Studies of Men and Books.* London: Chatto & Windus, 1895.

Stille, Alexander. "Library Privileges," *The New Yorker*, September 28, 1998: 43–46, 57–59.

Streeter, Burnett Hillman. *The Chained Library: A Survey of Four Centuries in the Evolution of the English Library.* London: Macmillan, 1931.

Thompson, Anthony. *Library Buildings of Britain and Europe: An International Study, with Examples Mainly from Britain and Some from Europe and Overseas.* London: Butterworths, 1963.

Thornton, Dora. *The Scholar in His Study: Ownership and Experience in Renaissance Italy.* New Haven: Yale University Press, 1997.

Vitruvius. *The Ten Books on Architecture.* Translated by Morris Hicky Morgan. New York: Dover, 1960.

Vogel, Steven. *Cats' Paws and Catapults: Mechanical Worlds of Nature and People.* New York: Norton, 1998.

Wagner, Patricia Jean. *The Bloomsbury Review Booklover's Guide: A Collection of Tips, Techniques, Anecdotes, Controversies & Suggestions for the Home Library.* Denver: The Bloomsbury Review, 1996.

Walter, Frank K. "Random Notes on Metal Book Stacks," *Library Journal* 53 (1928): 297–300.

Walters Art Gallery. *The History of Bookbinding: An Exhibition Held at the Baltimore Museum of Art, November 12, 1957 to January 12, 1958.* Baltimore: Trustees of the Walters Art Gallery: 1957.

Weinstein, Krystyna. *The Art of Medieval Manuscripts.* San Diego: Laurel Glen Publishing, 1997.

Wernick, Robert. "Books, Books, Books, My Lord!" *Smithsonian,* February 1998: 76–86.

Wheeler, Joseph L., and Alfred Morton Githens. *The American Public Library Building: Its Planning and Design with Special Reference to Its Administration and Service.* New York: Scribner's, 1941.

Williams, Joan. *The Chained Library at Hereford Cathedral.* Hereford, Eng.: Hereford Cathedral Enterprises, 1996.

Winterson, Jeanette. *Art & Lies: A Piece for Three Voices and a Bawd.* London: Vintage, 1995.

Woodsmith Magazine. *Bookshelves & Shelves.* Des Moines: August Home Publishing, 1996.

Wright, C. E. "The Dispersal of the Monastic Libraries and the Beginnings of the Anglo-Saxon Studies. Matthew Parker and His Circle: A Preliminary Study," *Transactions of the Cambridge Bibliographical Society* 1 (1949/53): 208–237.

Illustrations

PAGE:

Index

Italicized page numbers refer to figures and their captions.